The Gist of Genetics

The Gist of Genetics

Guide to Learning and Review

Rowland H. Davis
Stephen G. Weller

University of California, Irvine

Jones and Bartlett Publishers
Sudbury, Massachusetts

Boston   London   Singapore

Editorial, Sales, and Customer Service Offices
Jones and Bartlett Publishers
40 Tall Pine Drive
Sudbury, MA 01776
978-443-5000
info@jbpub.com
http://www.jbpub.com

Jones and Bartlett Publishers International
Barb House, Barb Mews
London, W6 7PA
UK

Library of Congress Cataloging-in-Publication Data

Davis, Rowland H.
    The gist of genetics : guide to learning and review / Rowland H. Davis,
Stephen G. Weller.
        p.   cm.
    Includes index
    ISBN 0-86720-919-4
    1. Genetics.   I. Weller, Stephen G.   II. Title.
QH430.D38 1996
575.1--dc20

95-40900
CIP

Executive Editor: Brian L. McKean
Marketing Manager: Rich Pirozzi
Project Editor: Kathryn Twombly
Senior Production Editor: Mary Hill
Cover Designer: Mark Rodriguez
Manufacturing Buyer: Jenna Sturgis
Editorial Production Service and Typesetting: A-R Editions
Printing and Binding: Braun-Brumfield
Cover Printing: Coral Graphics Services, Inc.

Printed in the United States of America
00   10 9 8 7 6 5 4

Contents

CHAPTER 6

CHAPTER 18
BACTERIAL AND BACTERIOPHAGE GENETICS

CHAPTER 19
REGULATION OF GENE ACTIVITY

Preface

In genetics, there is a body of knowledge that every student of the subject should know. *The Gist of Genetics* presents that body of knowledge with familiar examples, few embellishments, and methods of learning the subject quickly and easily. The book is modern in its organization, starting with the nature of the genetic material, connecting this knowledge with the phenomena of inheritance, and giving summaries of molecular, prokaryotic and biochemical genetics in the process. Its main emphasis is on genetic logic and rationales and the connection between formal genetics and gene function. It is addressed to readers of all levels, even those with no background in biochemistry or molecular biology.

We have written this book for three reasons. First, the modern, comprehensive genetics books now available have buried basic information in a welter of detail about all areas touched by genetics. Some texts have simply summarized the canonical aspects of genetics in a few initial chapters. As a result, most college genetics teachers do not use the majority of the material in a standard genetics text, and students have difficulty finding what they must know. *The Gist of Genetics* is designed as an underground resource for such students.

Second, those teachers who prefer to focus on the applications of genetics in their own sub-field will find this book supplies all the necessary information on the basic phenomenology of inheritance and genetic analysis. This book is especially suitable for such teachers and their students. Used as a supplementary guide, *The Gist of Genetics* will free the teacher to choose a more specialized text that serves his or her orientation. At the same time, our book gives students an alternate, reasonably full treatment of genetics in compact form.

Third, the bulk of *The Gist of Genetics* can serve as the sole text of a sophomore, one-quarter course in elementary genetics, as it has in our course at the University of California, Irvine.

Special features of *The Gist of Genetics* are the checklist approach to problem-solving; the tables organizing and summarizing large bodies of information for easy retention; the large number of problems, most class-tested and with answers; an extensive glossary; and simple drawings that help to visualize genetic phenomena. We have strived for a clear readable text that recognizes the difficulties with concepts and vocabulary that beginning Genetics students invariably encounter.

Parts of this book, as it developed, were condensed to form study guides (the second with John Carulli) for the second and third editions of Daniel Hartl's *Genetics,* also published by Jones and Bartlett. The *Gist of Genetics* therefore shares many of the study problems and figures, and some of the text with those guides. However, most of the text is entirely new, seeking a self-sufficiency that the study guides were not designed to have.

We thank the following geneticists for reviewing one or more drafts of this book, including the study guides from which it is expanded: Mark Sanders, University of California,

Davis; David Evans, The University of Guelph; Michael Goldmand, San Francisco State University, Philip Hastings, University of Alberta; John Carulli, Collaborative Genetics; Stuart Brody, University of Califirnia, San Diego; Kenneth Jones, California State University, Northridge; Bob Ivarie, University of Georgia. We also thank Arthur Bartlett for encouraging our earlier efforts in writing the study guides mentioned above.

Rowland H. Davis
Stephen G. Weller

December 1995

Chapter 1

REPLICATION AND DNA

Living things reproduce themselves accurately. The information needed to do so lies in DNA, the molecular form of genes. The structure of this molecule is suited to bearing information in a form that cells can duplicate.

A. The Simplest Form of Heredity: Asexual Reproduction

Reproduction of living things is so natural that many of us think little about it. If we do, we associate the idea with the propagation of microorganisms and cells, the similarity of parents and offspring, and the constancy of species. Thinking further, however, we realize that over long periods, species evolve and diversify; a child differs from both its parents; and a single cell, the fertilized egg, gives rise to many specialized cells in an adult organism. In the first case, we are dealing with evolution: changes in the genetic constitution of populations. In the second, we are dealing with sexual recombination of genes from two parents. In the third, we are dealing with differential expression of genes during development, even though the genes themselves remain constant in the cells of the adult. It is hard to begin the study of the accuracy of inheritance in the face of these additional phenomena. Genetics came of age only when scientists began to resolve these phenomena from one another and finally to study the replication and biochemistry of simple organisms.

The simplest form of heredity is asexual reproduction. The multiplication of bacteria, the propagation of fungi through asexual spores, and the propagation of plants and some invertebrates with cuttings or body parts are examples of asexual reproduction. The resulting organisms closely resemble the "parents," except for rare, heritable variants known as mutants. Mutants may be antibiotic-resistant bacteria, flowers with novel colors, or fungal strains with new nutritional requirements. **Mutants arise rarely and suddenly and have specific, heritable alterations.** Three inferences about heredity can be made from such observations.

- The complexity of organisms, even bacteria, requires a great deal of specific information. The small size of bacteria, fungal spores, and plant cells, which all can regenerate a complete organism, means that the information required by an organism exists in miniature form in single cells.

- Because organisms can propagate indefinitely, the information must replicate with great accuracy. This gives hereditary character to the information.

- The existence of mutants suggests two important things. First, the heritability of the altered character points to mistakes in replicating the genetic information of the organism—mistakes that are perpetuated, just like the original information. Second, the specificity of mutations suggests that hereditary information must be organized into discrete units that mutate independently. These units are **genes,** and the alterations in mutant organisms are called **mutations.**

What, then, is the chemical form of hereditary information—the genetic material?

B. DNA, The Genetic Material

Oswald T. Avery and his associates first showed, in the 1940s, that deoxyribonucleic acid (DNA) could carry genetic information. Avery worked with the *Pneumococcus* bacterium, now known as *Streptococcus pneumoniae.* Geneticists were slow to appreciate his work because they felt bacteria were not at all similar in their genetic character to more complex organisms. Moreover, owing to an erroneous conception of DNA structure, no one could see how it could carry information. Almost everyone assumed that genes were proteins because the chemical complexity of proteins seemed to require that genes be equally complex.

Normally, each cell of *Pneumococcus* has a polysaccharide sheath around it, which gives a slimy, smooth appearance to colonies on agar medium. The normal strain is called S, for smooth. Occasionally, mutants lacking the sheath appear in large populations of S cells. Their colonies are small and have a rough, irregular outline and surface. Cells of this R, for rough, strain reproduce as R cells. The mutant character is therefore heritable. This strain was not virulent and failed to kill mice when injected into them.

A curious finding was made in the late 1920s by a British doctor, Frederick Griffith. While neither heat-killed S cells nor living R cells would kill mice, if he injected them with a mixture of the two, the mice would die of a pneumonic infection. It appeared that the dead S cells had transformed the R cells. Avery pursued this phenomenon, and he was able to detect **transformation** in the test tube. Heat-killed S cells could transform R cells into live S cells in numbers far greater than mutation would account for.

The live S "transformants" grew into large S populations. When heat-killed, these populations could transform more R cells to S cells. R cells remained rough cells if dead S cells were not added, and dead R cells could not transform living R cells. The experiments showed that a piece of mutable, specific, and heritable information (for the polysaccharide sheath) could be transferred from dead S cells to R cells. In short, the "transforming activity" had the characteristics of genetic material.

Like all cells, bacterial cells contain small molecules and macromolecules. When Avery made an extract of the dead S cells—that is, taking only the soluble materials—the extract was active in transformation. Avery wished to determine what sort of molecule in this extract was responsible for transforming R cells. Small molecules were too small to bear much information, and Avery found by removing them from the extract, that they were not required. The candidates among the macromolecules were lipids, polysaccharides, proteins, and the two types of nucleic acids: RNA and DNA. Avery found that among soluble macromolecules from S cells, lipids were not essential for transformation.

Removal of polysaccharide from the lipid-free extract also left the transforming activity intact. This important observation showed that the polysaccharide in the sheath did not carry the information needed to make more polysaccharide sheath. **The needed information was encoded in another type of molecule.**

The extract now contained mainly protein and nucleic acids. The removal of protein did not reduce the transforming ability of the extract. This surprised most scientists because they thought only proteins were complex enough to encode genetic information. Avery then removed RNA from the preparation, using the enzyme RNAase that broke down RNA but not DNA. This had no effect. DNA remained. When Avery removed or digested DNA, transforming activity was lost. **Further work showed that pure DNA from S cells could transform R cells to the S type.**

Avery realized that DNA carried heritable information required for the formation of the polysaccharide sheath of *Pneumococcus*. Many years of further work by others showed that this was true for hereditary factors in all organisms. Avery's work stimulated further study of DNA to see how its structure could carry genetic information.

C. DNA Structure

DNA can serve as genetic material because of its unusual macromolecular structure. Let us consider the levels of structure in turn.

1. Nucleotides

Nucleotides are the repeating unit of DNA. A nucleotide has three components: phosphate, deoxyribose, and a nitrogenous base (Fig. 1–1). The phosphate is attached to the 5′ position of the deoxyribose. The nitrogenous base is attached to the 1′ position. Deoxyribose gets its name from the fact that the 2′ position lacks an oxygen atom that is found in ribose. The numbers of the carbon atoms of the sugar are primed to distinguish them from the ring atoms of the nitrogenous base.

Four kinds of nitrogenous base are found in DNA: adenine (A) and guanine (G), which are purines; and cytosine (C) and thymine (T), which are pyrimidines. A, G, C, and T are called nitrogenous bases because they contain several atoms of nitrogen and are chemically basic; that is, positively charged in neutral solution. Pyrimidines are one-ring structures; purines are two-ring structures. They attach to deoxyribose through one of their nitrogens (Fig. 1–2).

2. Nucleotide chains

Nucleotide chains are linear and unbranched. The phosphate of one nucleotide (attached to the 5′ position of its deoxyribose) is attached to the 3′ position of the deoxyribose of the next nucleotide. Thus a nucleotide chain has a sugar phosphate backbone with nitrogenous bases hanging from the deoxyriboses. Nucleotide chains may have millions of nucleotides with no chemical restriction on the sequence. **The nucleotide order, in fact, is the form of the information carried by DNA.**

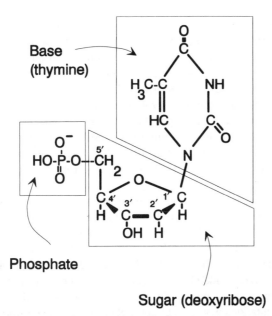

Figure 1–1   Structure of a nucleotide (thymidylic acid).

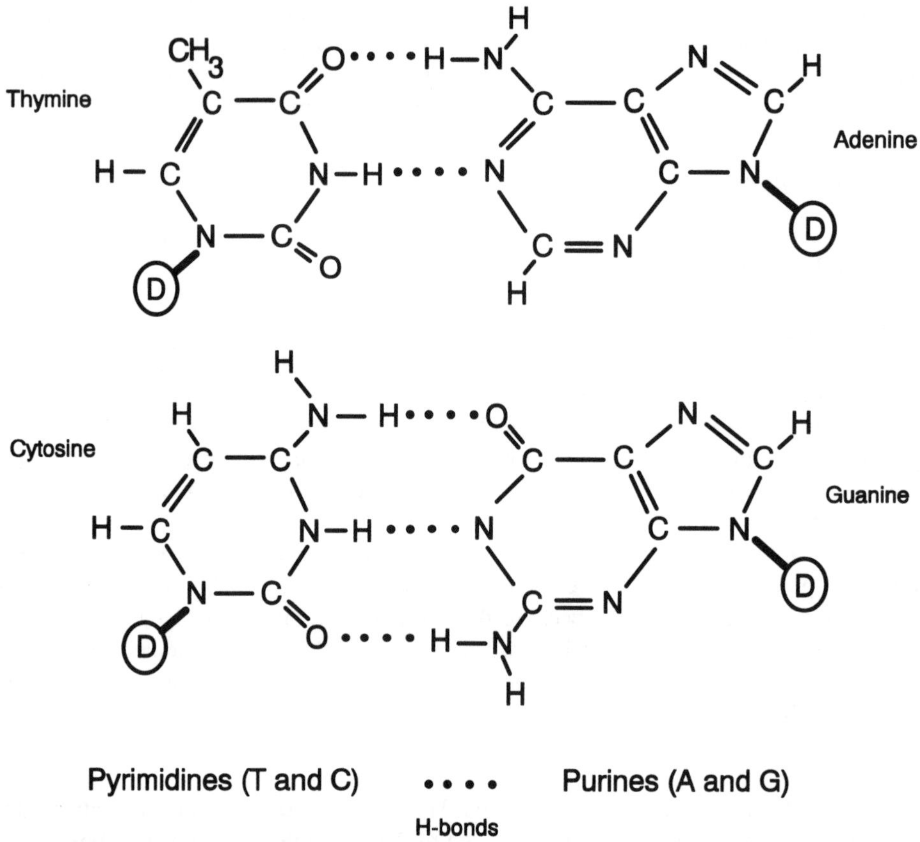

Figure 1–2   The base pairs of DNA, thymine-adenine (T-A) (above) and cytosine-guanine (C-G) (below). The complementary bases are connected by hydrogen bonds (dotted lines). Circled Ds represent deoxyribose, to which bases are attached.

3. The double helix

A DNA molecule has two chains, coiled around one another to form a double helix. At any point in the double molecule, a purine in one chain faces a pyrimidine in the other (Figs. 1–2 and 1–3). The only pairs found are adenine-thymine (A-T or T-A) and guanine-cytosine (G-C or C-G). These are called complementary pairs (Fig. 1–2). The restriction to these pairs is due to the ability of their members to form **hydrogen bonds,** which are shared hydrogen atoms that weakly bind the two chains together. Because the chains are so long, the many hydrogen bonds of nucleotide pairs (often called base pairs) hold DNA chains together at normal temperatures. The two nucleotide chains of a DNA molecule are antiparallel. Starting at one end, one chain goes from 5′ phosphate, through the sugars, to 3′ OH; whereas the other goes from 3′ OH to 5′ phosphate (Fig. 1–3).

The complementary relation of the chains of the double helix allows us to deduce the exact nucleotide sequence of one chain from the sequence of the other chain. Therefore, the DNA double helix contains all information in two forms.

Watson and Crick, who proposed this structure for DNA in 1953, saw (i) that the sequence of nucleotides could carry information (as a code); (ii) that replication might proceed by having each nucleotide chain determine the sequence of a new, complementary chain; (iii) that erroneous base-pairing during DNA synthesis could be the source of mutation; and (iv) that mutations would be as heritable as the original information. Their basic hypotheses were confirmed by the work of many other scientists since 1953.

A

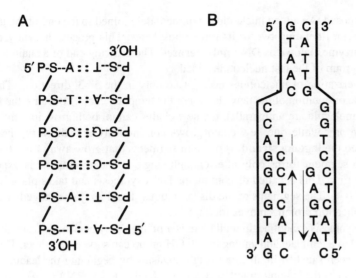

B

C

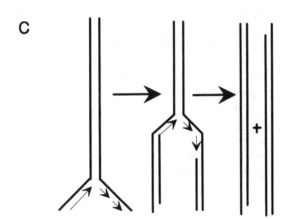

Figure 1–3 The antiparallel nature of DNA chains (top left) and the mode of DNA replication (top right and bottom). A new chain must be synthesized by additions to the 3'OH (top right); the two chains are locally synthesized in opposite directions, but the replication fork moves in one direction (bottom).

D. Semiconservative Replication

Cellular proteins contribute to DNA replication by properly positioning the DNA and catalyzing the required chemical reactions. DNA therefore does not really replicate itself; it contributes information for the synthesis of new nucleotide chains. Each of the two "parental" nucleotide chains serves as a **template** for the assembly of a new, complementary nucleotide chain. Overall, DNA replication is **semiconservative**: replication preserves the integrity of each half-molecule (nucleotide chain) of the parental DNA. In the daughter molecules, one chain (the template) is old DNA and the other is new.

Replication consists of (i) separation of nucleotide chains; (ii) positioning of free, complementary nucleotides against those of the parental chains; and (iii) polymerization (end-to-end attachment) of new nucleotides into new chains. All three processes happen continuously, with strand separation forming a replication fork (Fig. 1–3). Little single-stranded DNA is exposed at one time. One of the two new nucleotide chains grows into the replication fork as the parental strands unwind. Free nucleotides (as 5' triphosphates) make hydrogen bonds with their complementary nucleotides in the single-stranded template as they are exposed. An added nucleotide is immediately joined to the end of the growing nucleotide chain, liberating two of its three phosphates. This process is catalyzed continuously by enzymes known as DNA **polymerases.** The growing end of a chain is defined by the 3′ OH group of the last nucleotide added.

Polymerization of nucleotides takes place only in the 5′–3′ direction. (The direction refers to the orientation of the new chain, not the template chain.) Because the two strands of the old molecule are antiparallel, the new chains cannot both grow into the replication fork. Therefore, while one new chain grows continuously into the fork, the other new chain, called the **lagging strand,** is made in segments that grow away from the fork (Fig. 1–3). Each segment forms only after enough single-stranded template is exposed at the replication fork. This is called discontinuous DNA synthesis and takes place in all organisms (Fig. 1–4). The segments of the discontinuous strand are joined together as the complex of replication proteins moves forward.

DNA synthesis cannot begin until there is a **primer.** Any nucleotide chain that is hydrogen-bonded to a template and ending in a 3′ OH group can serve as a primer. The need for a primer has been satisfied in all free-living organisms by beginning the leading strand, and every segment of the lagging strand, with a short chain of RNA. RNA, as we will see later, is a single-stranded chain of ribonucleotides. The sugar carries an OH group lacking in deoxyribonucleotides on the 2′ position, and the nitrogenous base thymine is replaced by a similar base: uracil. RNA is made by various forms of RNA polymerase, which, unlike DNA polymerase, does not need a primer. The RNA segments used as primers in the lagging strand are made by an RNA polymerase known as **primase** (Fig. 1–4).

DNA polymerase has the ability to remove nucleotides—both ribo- and deoxyribonucleotides—from the end of another chain immediately in front of it, and replace them with deoxyribonucleotides as part of the chain it is itself making. As the replication fork moves on, the RNA nucleotides of the primers of each discontinuous segment are removed by the DNA polymerase that is making the next segment (nearer the fork), and are replaced with DNA nucleotides. These DNA segments, which abut one another, are finally ligated (i.e., connected) by an enzyme called **DNA ligase.** As a result, the lagging strand becomes continuous except in the region near the replication fork.

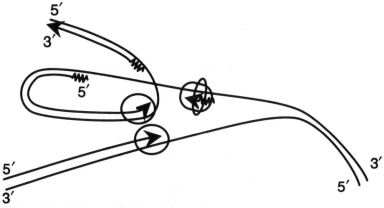

Figure 1–4   Conceptual organization of DNA synthesis near the replication fork. The leading strand (lower) grows into the fork as a DNA polymerase enzyme molecule (circle) adds nucleotides to its 3′OH end. The lagging strand, also with a DNA polymerase molecule extending it, moves away from the fork, following its template. As the parental strands separate (right) a new lagging-strand segment is initiated with an RNA primer (jagged lines). This RNA primer is made by a primase (oval) and extended thereafter with DNA nucleotides by another type of DNA polymerase (oblong). The RNA nucleotides of the primer will later be removed by the DNA polymerase (circle) making the segment behind it. The lagging-strand segments are joined into a continuous chain by DNA ligase (not shown).

Linear DNA molecules encounter a peculiar problem in replicating. Because DNA polymerase must have an RNA primer in order to begin, polymerization of deoxyribonucleotides (or even ribonucleotides) cannot begin at the 3′ end of the template. Unless there were a compensating mechanism for this problem, DNA molecules would become shorter and shorter each generation. Eukaryotes, having linear DNA, solve this problem with special repeated sequences called **telomeres,** found at the ends of chromosomes. Telomeres are special DNA sequences that are made separately by another enzyme, not by DNA polymerase. They are then added to the linear DNA molecules after replication to compensate for the incompleteness of replication by DNA polymerase.

The bacteria and their viruses, with circular DNA molecules, do not have to deal with this problem, because circular molecules lack ends. Circular molecules of DNA have two modes of replication. In one, called the theta mode, replication starts from a particular origin of replication at one point on the circular molecule and progresses in one or both directions around the circle (Fig. 1–5). In the other, called the **rolling circle** or sigma mode, one strand peels away from the circle creating a replication fork where it diverges (Fig. 1–5). Leading and lagging strand synthesis commence and continue, displacing a longer and longer tail originating with the divergent strand. The configuration maintains a double-stranded circle, which continuously "rotates" to provide template at the fork. In some small viruses only one strand, the circle, is copied; the product of the process is a longer and longer single-stranded tail peeled from the circle. Because the rolling-circle mode has no obligation to stop at one turn, the linear DNA may finally represent many tandem duplications of the information in the original circle.

E. Molecular Manipulation of DNA

DNA molecules are very long, and the sequences are astronomically variable. However, because there are only four bases, short specific sequences (four to ten bases) recur with predictable frequencies. Certain sequences characterized by their symmetrical nature (reading the same in both directions, if one considers the double-stranded molecule) can be recognized by **restriction enzymes.** These enzymes have evolved in bacteria to destroy foreign DNA, such as the DNA of bacterial viruses, that enters the cell. Restriction enzymes differ in the target DNA base sequences, called restriction sites, that they recognize and cut.

If a single restriction enzyme is used to cut a diverse population of molecules, all the fragments (except those arising from spontaneous breakage) will have the same ends; namely, one-half a restriction site. If DNA is isolated from a large population of cells and cut with a restriction enzyme, every fragment will be represented by many copies. Individual fragments, even those containing single genes, can be isolated in quantity by a variety of **cloning** methods. These methods rely on integrating fragments produced by one or more restriction enzymes into circular DNA **plasmids,** which are small circular DNAs that can replicate in bacteria (*see* Chapter 2). A number of methods allow molecular biologists to recognize bacteria carrying a particular plasmid with a gene of interest, to isolate it, and to make a pure preparation of the plasmid and the gene itself. These methods lie at the heart of the field known as **genetic engineering** and the industry known as biotechnology. Chapter 20 describes some of the basic tools and rationales of this field.

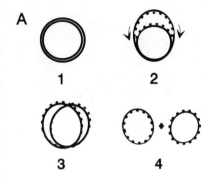

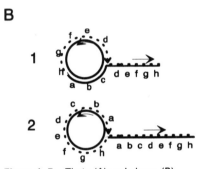

Figure 1–5 Theta (A) and sigma (B) modes of replication of circular DNA molecules.

REVIEW QUESTIONS

1. What attributes of DNA qualify it as the genetic material?

2. Starting with a single bacterial cell, how many cells will there be if division occurs once every 30 minutes for 5 hours?

3. In DNA, what are the nitrogenous bases attached to and in what positions?

4. What is the repeating unit of the *duplex* DNA molecule?

5. Why is there lagging-strand synthesis during DNA replication?

6. The difference between the code (information) for a characteristic and the characteristic itself is exemplified in Avery's transformation experiments. How?

7. What is a primer in DNA synthesis, and why is it needed?

8. In double-stranded DNA, how many times is the information present and in what forms?

9. Before 1950, most people thought that the genetic material had to be protein. Why did they think so, and what changed their minds?

10. Avery identified DNA as the material that could transform rough (R) *Pneumococci* to the smooth (S) phenotype. Why did he think this DNA might be genetic material?

11. Hydrogen bonds hold the two nucleotide chains of DNA together. What are the important features of these hydrogen bonds?

12. Explain the difference between the theta and sigma modes of replication.

13. What is a restriction enzyme, and how is it used?

14. The sequence recognized by a restriction enzyme, *Taq*I, is 5′ TGCA 3′ (the opposite strand has the same sequence but pairs in an antiparallel fashion to the one given). How frequently does this sequence appear in DNA, assuming that all bases are equally frequent?

Chapter 2

ASEXUAL REPRODUCTION AND DNA DISTRIBUTION

Bacteria, viruses, and cells of eukaryotes distribute their DNA to progeny in different ways during asexual reproduction. This is correlated with the form the DNA takes in these different types of organisms. Mitosis is the process in which eukaryotes distribute one copy of each DNA molecule to daughter cells during cell division.

A. Prokaryotes

Prokaryotes comprise the bacteria, including the blue-green algae. In prokaryotes, no membrane separates DNA from the rest of the cell. In fact, these organisms have none of the membranous organelles such as chloroplasts, mitochondria, vacuoles, Golgi apparatus, and endoplasmic reticulum, found in higher organisms.

In general, a single long circular DNA molecule carries the genetic information of a prokaryotic cell. Basic proteins associate only loosely with DNA in prokaryotes; the DNA is therefore more open and accessible than it is in eukaryotes. At one point on the circular DNA molecule, the DNA is attached to the bacterial cell membrane. Most of the DNA, however, is tangled up in a central **nuclear region.** Most people call the nuclear region a nucleoid, or even a nucleus, and the bacterial DNA molecule is often called a chromosome. These terms, as we will see, are borrowed from eukaryotes, but the structures of the elements differ in fundamental ways.

Replication of the double-helical circle starts at a point called the origin of replication. At this time, the point of attachment to the bacterial cell membrane becomes double, and from this point, two **replication forks** move in opposite directions around the circle (Fig. 2–1). The forks finally meet opposite the origin. The two daughter circles separate as they form, as their points of attachment to the membrane become farther apart. When the original circle of DNA is fully replicated, cell membrane and cell wall form between the daughter DNAs and divide the original cell into two daughter cells. Frequently the division of nuclei gets ahead of cell division, so that most cells of a population have two or four "nuclei."

Most prokaryotic species contain additional small circular molecules of DNA called plasmids. These are usually dispensable to viability, but they may, depending upon the particular plasmid, endow their hosts with additional capabilities such as the ability to mate, to neutralize antibiotics, or to make toxins that inhibit growth of competing bacteria.

B. Viruses

Viruses infect all types of organisms including bacteria, fungi, animals, and plants. Bacterial viruses are called **bacteriophages** or **phages.** Most bacterial viruses are made up only

A. Bacteria

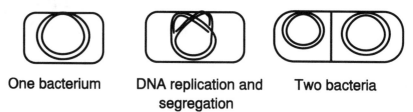

One bacterium   DNA replication and   Two bacteria
                   segregation

B. Viruses (bacterial)

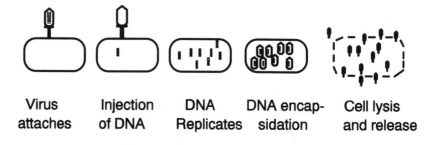

Virus   Injection   DNA   DNA encap-   Cell lysis
attaches   of DNA   Replicates   sidation   and release

C. Eukaryotic cells

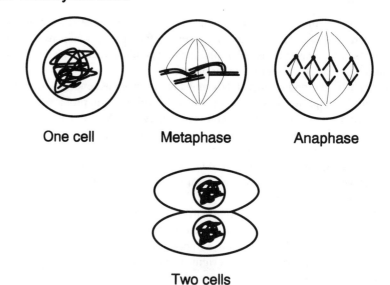

One cell   Metaphase   Anaphase

Two cells

Figure 2–1 Replication modes of DNA in bacteria, viruses, and eukaryotic cells.

of protein and nucleic acid. The protein forms a coat around the nucleic acid. Some bacterial viruses have a complex "tail" by which the phage attaches to the outside of the bacterium, and through which DNA is injected into the cell. Many viruses have RNA as their genetic material. RNA viruses represent the only exception to the rule that DNA carries the inherited genetic information. Finally, viruses of higher animals are often coated with some of the membrane of the cell in which they were formed.

Viruses are not free living, because they are incapable of metabolic activity, making proteins, or replicating their nucleic acids. They must enter living cells if they are to propagate (Fig. 2–1). Introducing only the nucleic acid into the host cell suffices to start an infection.

The nucleic acid carries information for (i) proteins that stop the functions of the host cell, (ii) proteins that help replicate the virus nucleic acid, (iii) some or all proteins of the virus coat, and in many cases, (iv) proteins that destroy the cell membrane or cell wall so that progeny viruses escape. In virus infections, DNA begins to replicate before coat proteins appear. The coats, when they appear, enclose condensed daughter DNA molecules and form mature virus particles called virions. The genetics of bacteriophages is discussed in Chapter 18.

Virus infections often kill cells, but many viruses multiply and escape through the membrane without killing the cell.

C. Eukaryotes

1. Nuclei and chromosomes

In eukaryotic cells, a nuclear membrane encloses the genetic material. The DNA of eukaryotes, unlike that of prokaryotes, consists of several to many *linear* molecules. The DNA molecules are strongly associated with basic (as opposed to acidic) proteins called **histones,** many other **nonhistone chromosomal proteins,** and a small amount of RNA. The aggregate of all these components is called **chromatin,** and each DNA molecule, together with its protein and RNA, is called a **chromosome.**

In all organisms and in higher mammals in particular, there is an acute "packaging" problem. **For instance, in every cell of humans there is over 1.5 meters of DNA.** When we consider that this DNA must replicate accurately in the same small space, we have difficulty understanding how such compaction can be maintained. One attribute of prokaryotic and eukaryotic DNA that makes it more compact is **supercoiling,** in which the helical molecule is twisted further like an over-wound telephone cord. In bacteria, domains of supercoiled DNA are defined by anchoring loops in some manner within the cell. In eukaryotes, another organization prevails: DNA is wrapped around octamers (a complex of 8 protein subunits) of the basic histones. The octamers consist of two molecules each of histone H2A, H2B, H3, and H4. DNA (about 120 base pairs) wraps around each octamer twice. The octamers greatly shorten the DNA molecule and impart a beaded quality to it. The "beads" are called **nucleosomes.** A molecule of histone H1 tightens the DNA around the octamers. The nucleosomes are linked to one another by a variable length of DNA averaging 55 base pairs. This beaded chromatin fiber can be further coiled into a regular solenoid, and this larger fiber is compacted again by two further orders of coiling. The compaction creates the chromosome fibers easily visible in the electron microscope and, in mitosis (see Fig. 2–4), the fully condensed chromosome.

In general, every eukaryotic organism has one (**haploid** organisms) or two (**diploid** organisms) complete sets of genes or **genomes.** Each set of genes is distributed among a number of chromosomes characteristic of each species. The bread mold *Neurospora crassa,* for instance, is haploid and has one set of seven chromosomes. A human, a diploid organism, has two sets of 23 chromosomes, while the diploid fruit fly *Drosophila melanogaster* has two sets of four chromosomes. Chromosomes vary greatly in length: the shortest in *Drosophila* has about 30,000 nucleotide pairs, and the longest (over 1 cm) in humans has over one hundred million nucleotide pairs.

As cells prepare to divide, the chromosomes condense and become microscopically visible. At this time, two other features of chromosomes appear: (i) Each chromosome at this stage consists of two identical copies, called **sister chromatids,** which have arisen by duplication of the original chromosome; (ii) They are held together at a constriction, the position of the **centromeres.** The centromeres of sister chromatids will be pulled apart during cell division, distributing one of the chromatids to each daughter cell (Figs. 2–1 and 2–2). Chromosomes of a set may vary both in their length and in the positions of their centromeres.

Each eukaryotic chromosome has two specialized structural features (Fig. 2–2). The centromere is one, having evolved as a point of attachment of spindle fibers during metaphase and anaphase of cell division. Centromeres have similar DNA sequences in any given organism, and in most eukaryotes **satellite DNA**—DNA of simple and highly

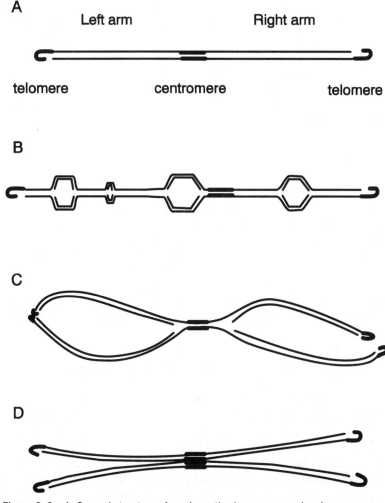

Figure 2–2   A: General structure of a eukaryotic chromosome, showing an arm on each side of the centromere, and the specialized telomeres at the ends. B, C, D: successive stages of replication of a chromosome, showing the multiple origins of replication, each moving outward such that the entire DNA molecule, including the centromeric DNA, is replicated by the end of the S phase.

repetitive sequence—is found close to the centromeres. The centromere is specialized for binding an aggregate of proteins known as the kinetochore, which in turn binds the spindle fibers. This is required for the orderly distribution of daughter chromosomes to daughter cells during mitosis. The other specialized structures are telomeres, one being found at each end of linear chromosomes, as we saw in Chapter 1. Telomeres are long multiples of four to six base pairs whose sequence is characteristic of the species. Telomeres are required for the complete replication of the chromosome, and they prevent chromosomes from joining together end to end. Telomeric DNA is constantly added to chromosomes as the number of repeated copies of the sequence diminishes through incomplete replication of the lagging strand of DNA.

2. The cell cycle

Cell growth, DNA replication, and the segregation of chromosomes to daughter cells are coordinated in the **cell cycle.** The cell cycle has four phases in continuously dividing cells (Fig. 2–3).

i. G_1 is a period ("gap") between the end of cell division and the beginning of DNA synthesis. During this period, considerable metabolic activity and cell growth occur.

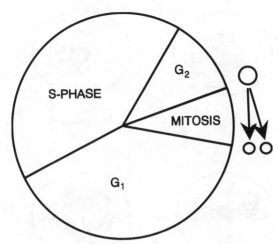

Figure 2–3 The cell cycle. The diagram is read clockwise, passing through G_1, DNA synthesis, and G_2, and culminating in mitosis, in which one cell becomes two.

ii. **S phase** is the period of DNA synthesis (S stands for synthesis). The cell synthesizes DNA at replications forks, described in Chapter 1. However, correlated with the larger amount of DNA in eukaryotic cells, each chromosome has multiple origins of replication whose replication forks diverge and ultimately fuse with others along the chromosome until the entire chromosome is replicated (Fig. 2–2). In this way sister chromatids, both having double-helical DNA molecules, are made from the original chromosome.

iii. G_2 is a second period or gap between S phase and cell division. This period is relatively short.

iv. In **M phase** or **mitosis,** chromosomes are distributed to daughter cells. You will notice that in the eukaryotic cell cycle, **DNA replication takes place at a different time than DNA distribution to daughter cells (Fig. 2–3).** This is a fundamental difference between division of eukaryotic and prokaryotic cells.

3. Mitosis

Mitosis (Fig. 2–4) starts with **prophase,** when chromosomes condense and the two sister chromatids become microscopically visible. The nuclear membrane disappears. The chromosomes then line up so that at **metaphase,** all centromeres lie near the middle of the cell. Spindle fibers originating at **centrioles** at the two ends of the cell attach to the kinetochores at the centromeres of the chromosomes. Other fibers run continuously from one centriole to the other.

Before the next step, **anaphase,** the centromeres holding the chromatids together become visibly divided, allowing sister chromatids to be separated. Spindle fibers pull sister chromatids to opposite poles of the cell. Each pole of the cell gets one copy of each chromosome. In the final phase, **telophase,** a nuclear membrane forms around the chromosomes now at the poles of the spindle. In the nuclei, the chromosomes become diffuse as the G_1 phase of the cell cycle begins.

The term **interphase** refers to all parts of the cell cycle except mitosis. During interphase, chromosomes cannot be discerned individually in most cells. Some chromatin remains heavily condensed even during interphase and appears even during mitosis somewhat more condensed than other parts of the chromosomes. The condensed chromatin, called **heterochromatin,** is relatively inactive in gene expression; active chromatin is called **euchromatin.** Examples of heterochromatin are satellite DNA, often concentrated around centromeres, the Y chromosome of male animals, and the inactive X chromosome of female mammals (*see* Chapter 10).

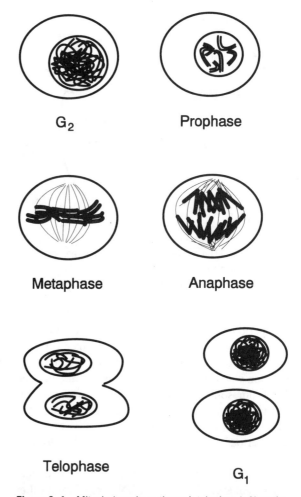

G₂ · Prophase · Metaphase · Anaphase · Telophase · G₁

Figure 2–4 Mitosis (prophase through telophase). Note the disappearance of the nuclear membrane after prophase, and the assembly of new nuclear membranes during telophase.

An unusual type of chromosome, first interpreted correctly in *Drosophila mela-nogaster,* is called a **polytene chromosome.** Such chromosomes appear in a number of organisms, but only in "terminally differentiated" cells that will not divide further. The large size of a polytene chromosome is due to the circa 1000 copies of the chromosome lying side by side to form a fat, banded structure. The many copies arise by **endoreduplication** (replication without separation of the daughter molecules), and the bands are due to the exact pairing of the same genes or sequences throughout.

D. Conclusion

All organisms divide and distribute their DNA to daughter cells with great accuracy. The process of DNA replication is similar in most organisms. DNA distribution varies; it takes place during replication in prokaryotes and after replication in eukaryotes. This is correlated with the existence of single circular DNAs in prokaryotes and multiple linear DNAs of eukaryotes. Viruses separate daughter DNA molecules by packaging each molecule in a protein coat.

The processes of replication and distribution of DNA represent heredity in its most fundamental form. Cells replicate their genetic information accurately, and daughter cells therefore closely resemble parental cells. These **uniparental** processes underlie asexual reproduction; reproduction **uncomplicated by differing contributions of two parents.**

REVIEW QUESTIONS

1. What problem does lagging-strand synthesis of DNA present to eukaryotic organisms that does not prevail in prokaryotes? How is this problem solved?

2. What is the function of the centromere? What is the equivalent mechanism in prokaryotes?

3. What is the structure of the nucleosome?

4. What are the characteristics of euchromatin, heterochromatin, and satellite DNA?

5. What is the temporal relationship between chromosome replication and mitosis in eukaryotic cells?

6. Distinguish between centromeres and centrioles.

7. In what ways do prokaryotes and eukaryotes differ?

8. We speak of bacterial DNA as a "chromosome," but how does it differ from the true chromosomes of eukaryotic organisms?

9. What is the difference between haploid and diploid cells?

Chapter 3

SEXUAL REPRODUCTION: LIFE CYCLES AND MEIOSIS

Sexual reproduction requires fertilization, the fusion of two gametes to form a zygote. Gametes carry one set of genes, and therefore zygotes carry two. Sexual reproduction also includes a process, called meiosis, that reduces the number of sets of genes from two to one. Meiosis is a sequence of two specialized nuclear divisions that is preceded by a single replication of DNA. The alternation of fertilization and meiosis is common to all eukaryotic organisms and constitutes the sexual life cycle. It permits two members of a species to contribute genes to their sexual offspring. The simplest genetic phenomenon in meiosis is the segregation of alternative forms of a single gene contributed by two parents during fertilization.

A. The Sexual Life Cycle

1. Sexual and asexual reproduction
In asexual reproduction of prokaryotes and eukaryotes, DNA simply replicates generation after generation. We now turn to sexual or biparental reproduction in which two parents contribute their genes to descendants. Most higher organisms incorporate biparental inheritance into their means of proliferation.

2. Haploids, diploids, and life cycles
Among eukaryotes, one sees haploid (1N) organisms, whose vegetative, or body cells have one set of chromosomes. The vegetative cells of diploids (2N), by contrast, have two sets. Each set of chromosomes is characteristic, in number and type, of the species and contains all the nuclear genes normally carried by that organism. Most fungi and certain algae are haploid; diploids comprise most of the plant world, protozoa, invertebrates, and vertebrates.

Even though organisms are referred to as haploid or diploid, all sexual organisms go through both haploid and diploid phases. In a sexual life cycle two events, **fertilization** and **meiosis,** alternate with one another. In fertilization, two haploid cells called **gametes** (usually from different parents) fuse to form a diploid **zygote.** In meiosis a specialized diploid germ cell, after replicating its DNA, undergoes two special cell divisions to yield four haploid meiotic products. A diploid cell cannot normally be a gamete, nor can a haploid cell undergo meiosis. Three basic patterns of the sexual life cycle exist in nature.

- **Haploid.** In haploids, haploid gametes are mitotic derivatives of the vegetative cells. The diploid phase, when formed, is immediately committed to meiosis (Fig. 3–1).

- **Diploid.** In diploids (such as animals and plants), meiosis takes place only in specialized germ cells such as oocytes or spermatocytes. The haploid phase of diploid organisms consists of the meiotic products which differentiate, with few or no intervening divisions, into gametes—eggs and sperm (animals) or certain nuclei of ovules and pollen (plants). These gametic cells are committed to fertilization (Fig. 3–1).

- **Haplo-diploid.** In some organisms (yeasts, mosses, ferns, and algae) both haploid and diploid vegetative states multiply by mitosis. In these organisms, the alternation of the two vegetative states is known as the **alternation of generations.**

Haploids, diploids, and organisms with both vegetative states may reproduce asexually by forming mitotic derivatives such as fungal conidia, fragmentation products,

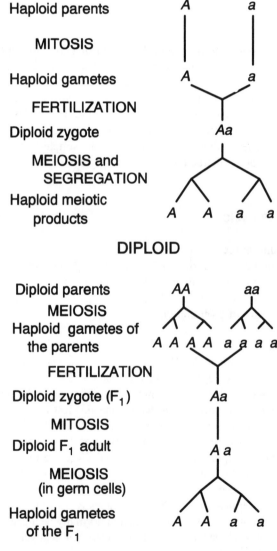

Figure 3–1 Haploid and diploid life cycles. The difference between them lies in the duration of the haploid and the diploid phases, but the sequence of events is the same.

zoospores and cysts, and successive life forms of parasites in successive hosts. The numbers of individuals increase in these asexual modes of reproduction, sometimes accompanied by the formation of cell types specialized for dormancy and dispersal. However, asexual reproduction leads to no genetic variation except through mutation.

B. Segregation in a Cross of Yeast Cells (*Saccharomyces cerevisiae*)

1. The life cycle of yeast

Yeasts are unicellular, and multiply by budding. *S. cerevisiae* has a haplo-diploid life cycle. Cells of both phases of the life cycle can proliferate indefinitely by mitosis.

Two haploid yeast cells, if they have different **mating types,** will fuse soon after they make contact, thereby accomplishing fertilization. The mating types, **a** and α, are determined by alternative forms of the mating type gene on one of the chromosomes. The diploid zygote (the product of fertilization) has both mating type determinants, but they are inactive, and no mating with other cells takes place in the diploid phase. As noted above, diploid yeast cells, like haploid cells, may reproduce indefinitely by mitosis.

Diploid yeast cells, when starved for nitrogen and carbon, stop growing and undergo meiosis. This process is preceded by one round of DNA replication and consists of two nuclear divisions. The net outcome is four **ascospores,** the meiotic products, enclosed in the cell wall of the original cell. Two spores will be mating type **a,** and two will be mating type α.

2. Thinking about a yeast cross

Let us cross haploid strains of yeast that differ by a mutation. Yeast colonies normally have a tan color. A mutation in the biochemical pathway that forms the nitrogenous base adenine causes the accumulation of a red pigment. When tan **a** cells are mixed with red α cells, pairs fuse to form diploid cells. The colonies formed by diploid cells by mitosis have a tan color. Moreover, they are unable to mate, indicating that the mating type gene is no longer active.

When the diploid cells of this mating are starved for nitrogen and carbon, each cell forms an ascus with four ascospores. **After germination two of the ascospores will grow into tan colonies, and two will grow into red colonies.**

A great deal has happened above, simple as it sounds. The events must be seen in formal terms in order to keep track of them.

- A "red" mutation is used, and its normal counterpart must be recognized when the two cell types are crossed. This brings up a need for gene symbols. Geneticists often name genes according to the **mutant** characters they give rise to. The alternative forms of these genes are called **alleles,** and they are often designated by modifying the gene name with superscript plus or minus signs, as in Figure 3–2, or by capital and small letters, as in Figure 3–1. In this example the red mutation might be called *red* or *red$^-$*, and the normal allele would be designated *red$^+$*. (The mating types of yeast are determined by alleles of a single gene, but neither is really a "mutation" of the other, and the nomenclature is unusual.)

- The diploid cells, the product of the mating, form tan colonies. This forces one to recognize the dominance relation between the *red* and *red$^+$* alleles in the diploid cell. We will explore this subject in later chapters; here we note that the tan determinant (*red$^+$*) is dominant to the recessive red determinant (*red*), since *red$^+$* masks the existence of *red* in the diploid cell.

- Sporulation produces four haploid cells from one diploid cell. The chromosomal events of this process, meiosis, lead not only to the **segregation** of the red and

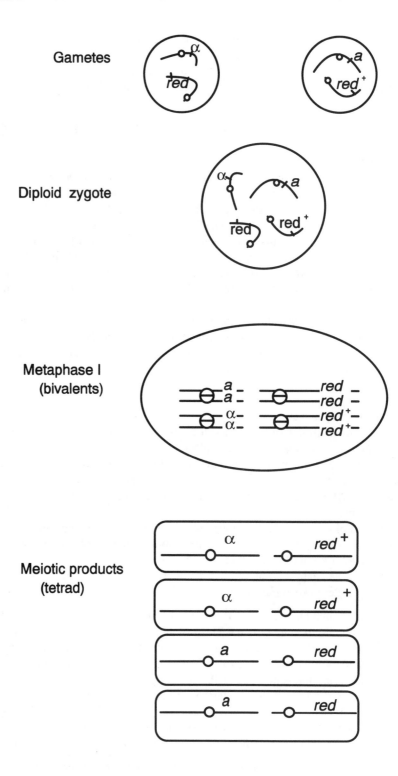

Figure 3–2 Summary of a yeast cross, starting with fusion of cells differing in mating type (**a**,α) and the red/tan colony color, determined by alleles *red* and *red*⁺. One of three possible tetrads is shown. Other possible tetrads will be discussed later.

tan determinants, unchanged from the original parents, but to their assortment with the mating type determinants. We will defer consideration of the latter process to a discussion of **independent assortment** in Chapter 5.

Meiosis and segregation are generalizable to all sexual eukaryotic organisms. Characters whose alternative states are determined by two alleles of a single gene yield 1:1 ratios among meiotic products. The regularity of the outcome of crosses involving alleles of a single gene (such as red^-/red^+ or mating types **a** and α) is enforced by meiosis. The 1:1 ratio of alternative characters among meiotic products is called **segregation.**

C. Meiosis and Segregation

Figure 3–2 depicts two (of 16) types of chromosomes in four successive stages of the yeast life cycle: (i) haploid gametes; (ii) the diploid cell before meiosis; (iii) metaphase of the first meiotic division; and (iv) the meiotic products. The two chromosomes in each haploid parental cell carry alleles of two different genes. One parent has the **a** mating type gene and the red^+ allele of another gene. The other parent has the α and red^- alleles of these genes, respectively. The events described below show how 1:1 ratios of alleles arise among meiotic products.

1. Fertilization and premeiotic S phase
Fertilization in yeast occurs only when the haploid cells that fuse are in the G_1 stage of the cell cycle. The diploid formed will therefore have two copies of each chromosome. The diploid then may undergo meiosis if it is induced to do so. The induction leads first to a replication of DNA. **A "premeiotic S phase" always precedes meiosis.**

2. Prophase I and Metaphase I
Meiosis starts with a complex prophase (Prophase I) that has little in common with prophase of mitosis. It will be described in detail in Chapter 9. However, a crucial event never seen in mitosis occurs here; namely, the pairing of **homologous chromosomes.** Homologous chromosomes, one contributed by each parent, carry the same genes though they may have alternative forms such as the mating type alleles. Homologous chromosomes pair in Prophase I, each member already having duplicated into two sister chromatids. Thus the outcome of chromosome replication in premeiotic S phase and the chromosome pairing in Prophase I yields the four-stranded configurations seen at Metaphase I. The four-strand structures are called **bivalents.**

During a short period in Prophase I the paired chromosomes undergo a breakage and reunion process known as **crossing over,** by which parts of homologous chromatids are exchanged. Crossing over is exact: no chromatid gains or loses any genes, but the process leads to the recombination of genes that lie on each side of the point of exchange on homologous chromosomes. This important process will be taken up in detail in Chapter 9. For our purposes here, crossing over is not relevant except for the possibility that segregation of alleles may be deferred to the second meiotic division if the chromatids of a bivalent exchange parts.

3. The meiotic divisions
So far, the cell has not divided (Fig. 3–2). **The rest of meiosis consists of two cell divisions without any further DNA replication.** The two divisions distribute the four chromatids (often called **"strands"**) of each bivalent to four different meiotic products. The first division separates the two homologous chromosomes by the two visible centromeres. The second division separates the sister chromatids of each chromosome after the sister centromeres become visibly doubled and physically separate. Overall, the diploid cell

yields four haploid meiotic products—each with only one set of chromosomes. **For any gene the meiotic products can be predicted from the bivalent; alleles of the same gene will have a 1:1 ratio among meiotic products (Fig. 3–2).**

All sexual organisms, haploid or diploid, undergo meiosis in this way.

D. Definitions and Summary

1. Chromosomes

The four-stranded entities formed by chromosome pairing in Prophase I are bivalents. The group of four meiotic products produced by a meiotic cell is called a **tetrad.** (Some developmental biologists call bivalents tetrads; we will not.)

The bivalent consists of two homologous chromosomes, or homologs. Each homolog has replicated to form two sister chromatids. In a diploid cell chromosomes come in pairs, one homolog from each parent. Homologs carry the same genes, though the form of the genes or alleles may be different (e.g., **a** vs. α).

Homologs recognize one another and pair only during meiosis. The pairing process in Prophase I is called **synapsis.** The pairing of homologs assures that each meiotic product of a tetrad gets one of each type of chromosome.

A chromosome may have one or two chromatids, which many geneticists call "strands." If the chromosome is two-stranded, each strand is called a **chromatid.** Sister chromatids are held together by a centromere. A chromatid becomes a chromosome when the centromere region holding the chromatids together becomes visibly doubled and the chromatids separate. Thus to count chromosomes, count centromeres.

Every chromatid has a double helix of DNA. That is the irreducible form of DNA in a chromosome or chromatid. *Do not be confused by the word "strand."* Unfortunately, it is used to designate both the two chromatids of a chromosome and the two nucleotide chains of a DNA molecule.

2. Meiosis and mitosis

In the last chapter, mitosis was described as a process that assured accurate duplication and equal distribution of DNA to the daughter cells. Cells that undergo mitosis can be either haploid or diploid. Mitosis of a haploid cell is pictured in Figure 2–4 (*see* Chapter 2). The diploid cell, unlike the haploid cell, is capable of undergoing meiosis as well as mitosis. However, in mitosis of a diploid cell, homologs do not pair. In Figure 3–3, mitosis and meiosis of a diploid cell having two pairs of chromosomes are compared. Try to grasp the essential differences in these processes by focusing on pairing and the behavior of centromeres, as well as on the number of cell divisions.

3. Genes and alleles

Alleles are alternative forms of genes. The word *gene* is therefore more general than *allele,* a specific form of a gene. Thus red^+ (on the short chromosome in Figure 3–2) may be a normal or wild-type allele, and red^- may be one of its defective mutant alleles. There are as many possible alleles of a gene as there are ways of mutating it. We recognize three relationships between alleles.

- **Historical.** One allele is derived from the other by mutation, or the alleles share a common ancestor.

- **Functional.** Alleles affect the same process: **a** and α control mating and red^+ and red^- may determine the presence or absence of an enzyme.

- **Positional.** Because alleles are related by mutation, they lie at the same position on homologous chromosomes, and are forced to pair during synapsis. Because alleles lie at the same point in a bivalent in Meiosis I, they are also forced to seg-

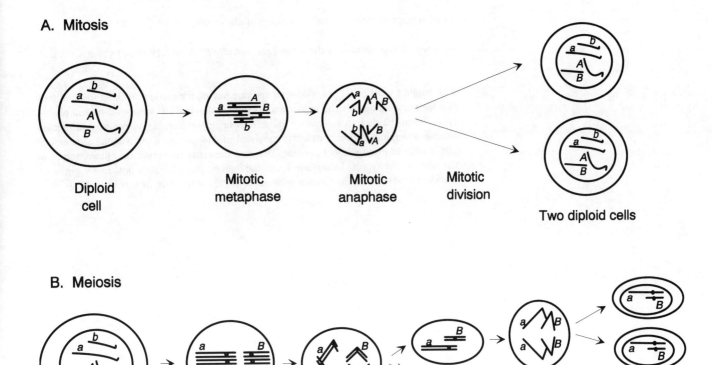

Figure 3–3   Mitosis (A) and meiosis (B) of a diploid cell having two pairs of homologous chromosomes.

regate from one another in the first anaphase (or the second, as we will see, if re-combination of chromosome parts occurs). **In no case can a normal meiotic product get both alleles or neither allele of a gene.**

REVIEW QUESTIONS

1. In what ways do haploid and diploid life cycles differ?

2. What are the stages and events in a sexual life cycle?

3. In what ways do mitosis and meiosis differ?

4. What is distinctive about Prophase I of meiosis?

5. How would you distinguish the terms bivalent, homolog, chromosome, and chromatid?

6. At what stage of meiosis does segregation of alleles take place?

7. How would you distinguish the terms *gene* and *allele?* What relationships characterize alleles?

8. Two haploid yeast cells, one wild-type and one having a mutation blocking the synthesis of adenine, were mated. The zygote and its mitotic descendants were wild-type in phenotype. When the diploid cells were induced to undergo meiosis, the progeny (a random sample of haploid meiotic products) consisted of 214 able to grow without adenine and 203 unable to grow without adenine. Diagram the cross (picturing the chromosomes of the parents, the zygote, Metaphase I, and the meiotic product), state how many genes are involved, and state the dominance relations of alleles of the gene or genes.

Chapter 4

SEGREGATION IN DIPLOIDS: F_1, F_2, AND TESTCROSSES

Because diploids have two sets of genes, genetic analysis of diploids requires at least two crosses. In the first, true-breeding parents (P) form a hybrid, the F_1. Germ cells of the hybrid which undergo meiosis yield haploid products in which segregation of alleles of genes contributed by the original parents has taken place. These haploid cells become gametes that must be used in a second cross to complete the analysis.

A. *Drosophila melanogaster*

Drosophila, the fruit fly, is a diploid insect. It has complex morphology with prominent red-brown eyes, short bristles, veined wings, and dark markings on its body. Mutations affecting these characters are easily recognized. The fly develops from an egg to a fertile adult in less than two weeks. Eggs hatch a day after they are laid. Larvae (small maggots) crawl about and eat for three or four days, molting three times. They then form a pupa enclosed in a dry case which in four or five days metamorphoses into an adult. The adult is fertile about 18–24 hours after emerging from the pupal case. *Drosophila* has been used widely to infer and study the principles of inheritance.

B. The Formation of Hybrid, F_1 Progeny from True-breeding Parents (P)

Crosses of diploid organisms like *Drosophila* are unexpectedly complicated, even when they involve only one gene. That is because (i) there are always two copies of every gene in the body cells (somatic, or vegetative cells) of the organism; (ii) **segregation of alternative characters cannot be observed** in haploid meiotic products, because they immediately become gametes; and (iii) alleles frequently display dominant-recessive relationships in hybrids in which one allele masks the presence of the other. Below, we point out these complications as we follow a one-gene cross of *Drosophila.*

1. Thinking about a *Drosophila* cross
Using symbols to represent crosses is the most important skill to develop early in your work in genetics. You must learn how to translate an innocent sentence describing the parents being crossed, for instance, into a symbolic representation that accounts for (i) each allele and (ii) each gene in (iii) both parents. Let us do this for a simple cross.

True-breeding, mutant, brown-eyed flies mate with true-breeding, normal flies, which have red-brown eyes. These flies are called the parental (P) generation. The eyes of the resulting **hybrids** are red-brown. These flies are called the **first filial generation,** or the F_1.

2. The checklist

Now represent the genetic constitutions of the flies symbolically, using a checklist:

a. What are the alleles and their symbols?   The parents differ in one gene that determines eye color. The *brown* mutation is allelic to the normal red-brown allele. In *Drosophila,* we designate both alleles with a symbol based on the mutant character. The mutation is *bw* and the normal allele is bw^+.

b. How many copies of the gene does the diploid have?   There must be two copies of every gene in each parent. Because the parents are "true-breeding" or pure, the two copies must be the same. The brown parent is *bw/bw,* and the normal parent is bw^+/bw^+.

c. What are the haploid gametes of the parents?   The brown and red-brown parents form haploid gametes (containing only one set of genes) by meiosis before the mating takes place. The gametes of brown flies are *bw,* and those of the red-brown flies are bw^+.

d. Does the hybrid contain the sum of parental gametes?   The gametes (sperm from the male and eggs from the female) will fuse to form zygotes. The zygotes will grow and develop, becoming the hybrid, diploid offspring of the parents. The hybrids have the constitution bw/bw^+.

e. From the appearance of the hybrid, what allele is dominant?   The only part not predictable about the cross in advance is the eye color of the hybrid flies. They are red-brown. **This shows that bw^+ is dominant over *bw*.**

f. What ratio of gametes do the F_1 flies produce when their germ cells undergo meiosis?   When the hybrid forms gametes they will be of two types, *bw* and bw^+, in equal frequency. This is what is expected of alleles of one gene in meiosis: a 1:1 segregation.

The checklist is a way of formulating a hypothesis about a problem in which the genetic basis of the characters is not yet clear. Its virtue, as the problems illustrate, is that if information given in a problem conflicts with the progress of the checklist, you are automatically forced to alter the hypothesis or the approach accordingly. **The checklist should be run through systematically in every cross until it is instinctive.**

3. Definitions

The **genotype** of a cell or organism is its genetic constitution. The diploid character of *Drosophila* requires two copies of every gene of interest in genotype notations. The genotypes in the above cross are bw^+/bw^+, *bw/bw* and bw^+/bw, the last being the hybrid. Genotypes, as we go on, may or not use the slash (/) symbol, but you can think of this symbol as separating homologous chromosomes with maternal and paternal contributions (or the reverse) as numerator and denominator. The phenomenon of linkage, however, will require that this "fractional" notation be used.

The **phenotype** of an organism designates its appearance or behavior. In our cross, red-brown eyes and brown eyes are phenotypes. Because of dominance, two genotypes (e.g., bw^+/bw^+ and bw^+/bw) may have the same phenotype. This phenotype can be represented as "bw^+ _," where the blank may be occupied by the dominant or the recessive allele.

With respect to a given gene, a **homozygote** is a diploid having identical alleles. The bw^+/bw^+ and *bw/bw* genotypes are homozygous for the "bw" gene. A **heterozygote,** like the bw^+/bw genotype, is a diploid having different alleles of a gene. The heterozygote is sometimes called a **carrier** of the recessive gene, because the organism carries it without revealing it in the phenotype. The loose term "hybrid" refers to the offspring of visibly different parents whose genotypes may not be known with certainty.

Genetic symbols vary among organisms. Table 4–1 lists the wild-type and mutant alleles of selected genes in several eukaryotes. Note how the conventions differ, and be prepared for the variety as we proceed.

C. Genetic Analysis of the F_1 Hybrid: the F_2, Backcrosses and Testcrosses

The gametes of the F_1 hybrid yield another generation of diploids only after fusing with other gametes. The mating may be with another bw^+/bw individual to form an F_2. It may be

Table 4–1 Gene symbols in various organisms

Organism	Wild-type allele	Phenotype	Mutant allele	Phenotype
E. coli	$hisA^+$	His$^+$ (histidine-independent)	$hisA$	His$^-$ (histidine-dependent)
Yeast	HIS2	His$^+$	his2	His$^-$
Neurospora	his-1$^+$	His$^+$	his-1$^-$	His$^-$
Drosophila	bw^+	red-brown eye	bw	brown
	Cy^+	normal wing	Cy (dom.)	curly wing
Maize	V	green shoot	v	virescent
Snapdragon	R	red flower	r	white
Pea	W	round pea	w	wrinkled
Mouse	a^+ or A	agouti coat	a	non-agouti
	bn	straight tail	Bn (dom.)	bent tail
Human	Hb^A	normal hemoglobin	Hb^S	sickle cell trait (Hb^A/Hb^S)
				Sickle cell anemia (Hb^S/Hb^S)
	H	non-hemophilic	h	hemophilia

a "backcross" to a bw^+/bw^+ homozygote. It may be a "backcross" to a bw/bw homozygote which for reasons given below is called a **testcross**.

1. The F_2 (bw^+/bw × bw^+/bw)

In this case, each parent contributes 1/2 bw^+ and 1/2 bw gametes. Random combination of gametes is best seen in the chart below yielding the diploid zygotes within the closed boxes:

Gametes of the male F_1

	1/2 bw^+ sperm	1/2 bw sperm
1/2 bw^+ eggs	1/4 bw^+/bw^+	1/4 bw^+/bw
1/2 bw eggs	1/4 bw/bw^+	1/4 bw/bw

Gametes of the female F_1

The sum of the zygotes is 1/4 bw^+/bw^+ + 1/2 bw^+/bw + 1/4 bw/bw. The zygotes grow to adult F_2 flies. Because bw^+ is dominant over bw, the phenotypic ratio is 3 red-brown : 1 brown. F_2s present three difficulties:

- Segregation occurs in the gametes of both parents;

- **Dominance** obscures the equality of alleles in the cross; and

- **Heterozygous flies can form in two ways:** bw^+ eggs + bw sperm, and bw eggs + bw^+ sperm. The arrangement of gametes on two sides of a chart makes these possibilities clear.

2. Backcross to dominant parent (bw^+/bw × bw^+/bw^+)

In this cross, the homozygous parent can produce only one kind of gamete: bw^+. The F_1 parent, as above, produces 1/2 bw^+ and 1/2 bw gametes. The result of the cross of bw^+/bw

$\times$ bw^+/bw^+ is 1/2 bw^+/bw^+ + 1/2 bw^+/bw. Because bw^+ is dominant, the phenotypes of the offspring are indistinguishable.

3. Backcross to recessive parent ($bw^+/bw \times bw/bw$)

The offspring of the cross of $bw^+/bw \times bw/bw$ is 1/2 bw^+/bw + 1/2 bw/bw. In this case, the ratio will be visible phenotypically. This type of cross is so useful that it is called a **testcross.** Testcross progeny can reveal the gametic ratio of a parent of unknown genotype.

D. About Testcrosses

A testcross is a mating of an unknown genotype with a "tester" parent that is homozygous recessive for any gene of interest. Its power lies in the inability of alleles from the tester to mask the segregating alleles of the unknown. For instance, if the red-brown flies (bw^+/bw^+ and bw^+/bw) of the F_2 were *individually* testcrossed, bw^+/bw^+ flies would produce only red-brown progeny and bw^+/bw flies would produce red-brown and brown progeny in a 1:1 ratio. That is how one would distinguish two genotypically different, but phenotypically identical flies.

The testcross progeny of a heterozygous diploid is formally the same as the meiotic progeny (1N) of a diploid yeast cell. In the latter case, however, one can observe the phenotypes of the meiotic products (actually the colonies derived from them by mitosis) directly. Moreover, in haploid cells, there is only one allele of each gene. Therefore, dominance does not complicate the relationship between phenotypes and genotypes in the haploid phase of organisms with haploid or haplo-diploid life cycles.

E. The Word "Segregation"

The word segregation always refers to alleles of a single gene. It is used by geneticists in three somewhat different contexts:

- The 1:1 segregation of **meiotic products** (or gametes) of a heterozygous cell,

- The 1:1 segregation of the diploid testcross progeny of a heterozygous diploid (which is equivalent to the ratio of gametes in the heterozygote), and

- The segregation of progeny phenotypes in an F_2 generation, such as the 3:1 ratio in the cross involving red-brown and brown.

F. Summary of Diploid Crosses Involving One Gene

Table 4–2   One-gene crosses

Parental genotypes	Progeny genotypes
$AA \times AA$	all AA
$AA \times Aa$	1 AA:1 Aa
$AA \times aa$	all Aa
$Aa \times Aa$	1 AA:2 Aa:1 aa
$Aa \times aa$	1 Aa:1 aa
$aa \times aa$	all aa

Table 4–2 lists all possible matings involving one gene. It is useful to be able to recognize at a glance (i) the outcome of any cross and (ii) the parents of any of the groups of offspring. Your mental image of the crosses should include the gametes of the parents as they combine to form the progeny listed.

REVIEW QUESTIONS AND PROBLEMS

1.   Two normal-looking *Drosophila* were mated and produced 45 normal and 13 ebony-bodied flies. How would you interpret this result?

2.   Black fur in guinea pigs is a dominant trait, whereas white fur is an alternative, allelic, recessive trait. In a series of crosses beginning with a pure-breeding black guinea pig and a white one, what fraction of the *black* F_2 (not the F_1) is expected to be heterozygous?

3.   A mating of a short-tailed mouse with a normal true-breeding mouse produces a 1:1 ratio of short-tailed to normal mice in the progeny. A mating of two short-tailed mice gives a 2:1 ratio of short-tailed to normal progeny. How can these results be explained?

4.   A breeder of Irish setters has a particularly valuable show dog but she knows it is descended from the famous bitch Rheona Didona, which was known to have carried a harmful recessive gene for atrophy of the retina. Before she puts the dog to stud, she must be sure that it does not carry this allele. How should she do this?

5.   In a testcross of a pea plant heterozygous for the recessive allele, *wrinkled,* what fraction of the progeny would be wrinkled?

6.   In diploids, what crosses allow you to observe segregation of alleles phenotypically?

Chapter 5

INDEPENDENT ASSORTMENT

Cells heterozygous for two genes (*AaBb*) on non-homologous chromosomes produce equal numbers of four kinds of meiotic products (*AB, Ab, aB,* and *ab*). In diploids, this will yield a 1:1:1:1 ratio in a testcross. Two doubly heterozygous parents will give rise to a progeny with a 9:3:3:1 phenotypic ratio (the doubly dominant category being the largest) if both genes have a dominant-recessive relationship. This complex ratio is nothing more than a combination of two 3:1 ratios, and one can easily predict it if the genetic basis of the traits is known. However, the origin of genetic combinations, based in part on chance, requires that objective statistical tests be used to compare expectations of a cross with its actual outcome.

A. Overview

Genes lying on non-homologous chromosomes are said to be unlinked. Independent assortment is the simultaneous segregation of alleles of two or more unlinked genes. Genes may also assort independently if they are on homologous chromosomes but so far apart that they recombine freely in meiosis. We will discuss the latter case in Chapter 9.

Meiotic products of haploid and diploid organisms appear in a 1:1 ratio for any pair of alleles. When we consider two genes, which have alleles *A* and *a* in one case and *B* and *b* in the other, the two 1:1 segregations yield four combinations in equal proportions: 1/4 *AB* : 1/4 *Ab* : 1/4 *aB* : 1/4 *ab*. Each meiotic product gets exactly one allele of each gene, never two or none. In haploids, the phenotypes correspond to the genotypes of these meiotic products. In diploids, a testcross of a parent heterozygous for two unlinked genes (i.e., *AaBb* × *aabb*) yields equal numbers of four genotypes, *AaBb, Aabb, aaBb,* and *aabb,* phenotypically revealing the ratio of the meiotic products (gametes) of the heterozygote.

In the F_2 of diploids, two alleles with a dominant-recessive relationship segregate phenotypically in a ratio 3/4 dominant (*A_*) : 1/4 recessive (*aa*). Notice that the dominant phenotypes, *AA* and *Aa,* are grouped together as "*A_*." If both parents are doubly heterozygous (*AaBb* × *AaBb*), the outcome is the more complex 9 *A_B_* : 3 *A_bb* : 3 *aaB_* : 1 *aabb* phenotypic ratio. However, this is nothing more than two 3:1 ratios; there are 12 *A_* : 4 *aa* and 12 *B_* : 4 *bb*. The genotypic ratio underlying the 9:3:3:1 phenotypic ratio is more complex, but easy to understand when we come to it.

B. P, F_1 and F_2 Generations of a Dihybrid Cross

Gregor Mendel, working in the mid-nineteenth century with inbred varieties of peas, developed the first clear rules of inheritance in diploid organisms. Because peas were

hermaphroditic, farmers could maintain pure lines of a variety of types by simply allowing flowers to self-fertilize. Mendel performed controlled pollinations by removing
anthers from flowers before they were mature, and adding pollen from another plant to
the pistil. Among his crosses was one between true-breeding plants producing smooth,
yellow peas and another producing wrinkled, green peas. Before we proceed, notice
that the two parents differ in two pairs of alternative traits (for surface texture and
color); possibly representing two pairs of alleles. Using your checklist (*see* Chapter 4),
try to formulate the cross with appropriate symbols.

All progeny (the F_1) of the cross, represented by the peas in the pods of the parental
female plant that was fertilized, were smooth and yellow. If one gene controlled each
trait, smooth must be dominant to wrinkled, and yellow must be dominant to green. At
the very beginning of the science of genetics, however, Mendel initially had no evidence
that the information for the wrinkled and green traits still existed in the F_1 progeny. His
first major contribution was to infer after a further cross that this information still existed,
unaltered, in the mature pea plants derived from the smooth, yellow F_1 peas.

When allowed to self-fertilize, the F_1 plants derived from these peas produced progeny peas in the following numbers:

315 smooth, yellow

108 smooth, green

101 wrinkled, yellow

32 wrinkled, green

The progeny can be resolved into two segregations: (i) 423 smooth : 133 wrinkled
(approximately 3:1) and (ii) 416 yellow : 140 green (approximately 3:1). **This illustrates
an indispensable mental step: resolving complex data initially into simpler components.** So far, the data meet the criterion (3:1 segregation) for one gene and two alleles
for each pair of alternative characters. Mendel inferred that the two F_1 plants, if they contributed equally to the offspring, must each have a copy of the recessive allele for each
gene. He in fact showed that the original parents could be reversed with respect to the
characters: the wrinkled, green parent could be represented by either the pollen or the
ovules, but the outcome was the same.

In the data above, two simultaneous 3:1 segregations occur. Because we look at them
separately, we have not yet tested whether the 3:1 distribution of one character has any
relationship with the phenotypic categories of the other. If we want to know this, how can
we formulate the question? Here is how: **If one 3:1 distribution occurs independently
of the other, then within each phenotypic category for one gene, the other gene
should segregate 3:1.** This is the independence of independent assortment.

Among the 423 smooth peas, the ratio of yellow and green is 315:108, which is approximately a 3:1 ratio. Among the 133 wrinkled peas, the ratio of yellow and green is
101:32, or again about 3:1. Therefore, yellow and green segregated **independently** of
smooth and wrinkled. The easiest way to see this diagrammatically is to make a tree of
classes and multiply the fractions of each class to get the composite phenotypes.

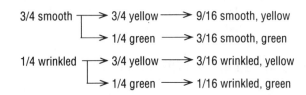

The criterion of independent assortment seen here (the 9:3:3:1 ratio) applies to two genes of a diploid in the F_2, each gene having dominant and recessive alleles.

C. What Happens in Meiosis During Independent Assortment?

Figure 5–1 uses an abbreviated picture of the life cycle to depict (i) true-breeding parents; (ii) parental gametes; (iii) the F_1 before meiosis; (iv) the Metaphase I configuration in the F_1; and (v) the meiotic products of the F_1.

The most important part of the picture is **the need to diagram two possibilities in Metaphase I.** The bivalents can line up in two equally probable ways, each having a different outcome. One alignment yields two *AB* products and two *ab* products, which we call **parental:** they have the same genotype as the parental gametes. The other alignment yields two *Ab* and two *aB* gametes, which we call **recombinant:** the alleles of the *A* and *B* genes are recombined in comparison to the parental gametes.

Because the two outcomes (Figure 5–1) happen equally frequently, meiotic products have a ratio of 1 *AB* : 1 *Ab* : 1 *aB* : 1 *ab*. **That is the criterion of independent assortment during meiosis.**

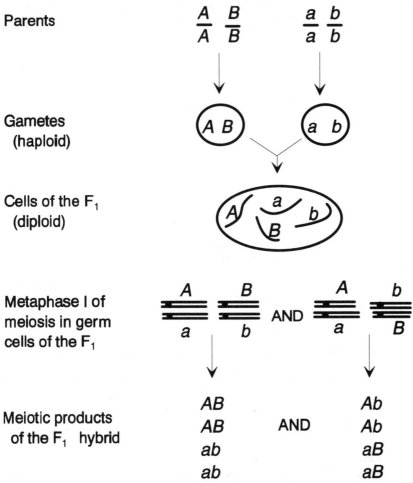

Figure 5–1 Formation of a dihybrid and the two Metaphase I orientations of independent bivalents of the dihybrid. The meiotic products are equal in frequency.

D. Haploid Crosses and Diploid Testcrosses Both Yield the 1:1:1:1 Ratio

1. Haploids

In haploid organisms, equal frequencies of meiotic products emerge from crosses involving independently assorting genes. We would diagram a cross of a haploid in this way:

			I	**II**
A B			*A B*	*A b*
×	$\dfrac{A}{a}\ \dfrac{B}{b}$		*A B*	*A b*
a b			*a b*	*a B*
			a b	*a B*

Parents	Transitory diploid	Meiotic products (of two meiotic cells with different alignments of bivalents) (Sum: 1/4 *AB*, 1/4 *Ab*, 1/4 *aB*, 1/4 *ab*)

The meiotic products will grow into mature haploid organisms (the vegetative phase of the life cycle) whose phenotypes are directly visible.

2. Diploids

In diploids, a testcross reveals the same phenomenon. The cross *AaBb* × *aabb* is a testcross of a dihybrid. The dihybrid produces gametes in the ratio 1 *AB* : 1 *Ab* : 1 *aB* : 1 *ab*. If each one is combined with a gamete *ab* (the only gamete the *aabb* tester can form), four genotypes with distinct phenotypes (in parenthesis) emerge in equal numbers from the cross.

1/4 *AaBb*   (*A_B_*)

1/4 *Aabb*   (*A_bb*)

1/4 *aaBb*   (*aaB_*)

1/4 *aabb*   (*aabb*)

The phenotypic designations carry forward a convention that implies ambiguity with the "blanks." In this case, however, there is no doubt that all blanks are occupied by recessive alleles.

E. The Genotypes Underlying the F$_2$ 9:3:3:1 Ratio (F$_1$ × F$_1$ Cross)

You have three ways of deriving the genotypes of the F$_2$. The first is called the **Punnett square;** it is easy to picture and hard to work with. The second is a **probability method;** harder to picture but extremely easy to work with. The third method is **memorization;** fast and dependable if you understand the probability method.

1. The Punnett square

This is a grid with the four gametic genotypes of the F$_1$s on the vertical and on the horizontal. The genotypes resulting from fertilization events—the F$_2$ generation—appear in the boxes.

	AB	Ab	aB	ab
AB	AABB	AABb	AaBB	**AaBb**
Ab	AABb	AAbb	**AaBb**	Aabb
aB	AaBB	**AaBb**	aaBB	aaBb
ab	**AaBb**	Aabb	aaBb	aabb

Sixteen equally possible combinations of gametes are shown, but there are only nine different genotypes. You could list all genotypes and their frequencies, but you will find it hard to be sure all categories are accounted for and summed properly.

The Punnett square, however, has an extremely informative pattern: (i) There are four different ways of getting a double heterozygote (boldface). (ii) There is only one way of getting each of the four double homozygotes (underlined) (iii) There are two ways of getting each of the four single heterozygotes (*AABb, AaBB, aaBb* and *Aabb*). We will see how this pattern arises in the probability method.

2. The probability method

The *A* gene yields three genotypes in the F_2 (1/4 *AA*, 1/2 *Aa* and 1/4 *aa*), as does the *B* gene. Independent assortment means that the same distribution of *B* genotypes is found within each *A* genotype, and that there are nine composite genotypes (3 for $A \times 3$ for *B*). For instance, among *AA*'s we will find *AABB, AABb,* and *AAbb*. To get the actual fractions of each composite category, simply multiply the frequencies of the components: among the total progeny, the *AABb* composite genotype is expected to appear in 1/4 (for *AA*) $\times$ 1/2 (for *Bb*), or 1/8 (2/16) of the entire progeny. Similar calculations give us the fractions of the other eight genotypes. **The principle used is that the probability of a composite genotype is the product of the probabilities of each genotypic component—once you know these probabilities are independent of one another.** We will return to this rule in a more general form later.

3. Memorization

Memorizing a systematic list of genotypes and their probabilities is extremely useful. It emerges naturally from the probability method. F_2 genotypes and their ratios (in 16ths) are listed below, with doubly dominant phenotypes first, single recessives next, and the double recessive last. You can assign ratios to all nine categories if you remember that double homozygotes are 1/16, double heterozygotes are 4/16, and genotypes with one heterozygous gene and one homozygous gene are 2/16, as we saw in the Punnett square.

	Genotypes		Phenotypes	
	(in 16ths)			
1	*AABB*			
2	*AABb*			
2	*AaBB*			
4	*AaBb*		9/16	*A_B_*
1	*AAbb*			
2	*Aabb*		3/16	*A_bb*
1	*aaBB*			
2	*aaBb*		3/16	*aaB_*
1	*aabb*		1/16	*aabb*

F. Many Independently Assorting Genes

The probability method gives you the power to predict any phenotype or genotype of any cross if dominance relations are known and the genes assort independently. What, for instance, is the fraction of *AaBbcc* progeny of the cross *AaBbCc* × *aaBbCc?* The trick is to **resolve the problem into one-gene components.** (The ratios for all one-gene crosses are given in Table 4–2, Chapter 4.) You know that an *Aa* × *aa* mating yields 1/2 *Aa* progeny; a *Bb* × *Bb* mating yields 1/2 *Bb* progeny; and a *Cc* × *Cc* mating yields 1/4 *cc* progeny. Therefore, the mating given will produce $1/2 \times 1/2 \times 1/4 = 1/16$ *AaBbcc* progeny.

Another problem: What proportion of the progeny of a *AaBbcc* × *aaBbCc* mating will have the phenotype *A_B_cc?* It will be $1/2 \ (A\_) \times 3/4 \ (B\_) \times 1/2 \ (cc) = 3/16$.

G. Diagnostic Ratios for Mono- and Dihybrid Crosses

Table 5–1 gives diagnostic characteristics of one- and two-gene crosses, the latter with independently assorting genes.

Table 5–1 Diagnostic crosses of monohybrids (*Aa*) or dihybrids (*AaBb*)

Cross or stage	Monohybrid (*Aa*)	Dihybrid (*AaBb*)
Gametes	1 *A*:1 *a*	1 *AB*:1 *Ab*:1 *aB*:1 *ab*
Genotype ratio in testcross (F₁ × *aa* or *aabb*)	1 *Aa*:1 *aa*	1 *AaBb*:1 *Aabb*:1 *aaBb*:1 *aabb*
Genotype ratio in F₂ (F₁ × F₁)	1 *AA*:2 *Aa*:1 *aa*	1 *AABB*:2 *AaBB*:2 *AABb*:4 *AaBb*: 1 *aaBB*:2 *aaBb*:1 *AAbb*:2 *Aabb*: 1 *aabb*
Phenotype ratio in F₂	3 *A_*:1 *aa*	9 *A_B_*:3 *A_bb*:3 *aaB_*:1 *aabb*

H. Probability and Genetic Analysis

By this time, the simpler rules of gene transmission should be clear, but the practical ways of formulating genetic expectations and testing them may be obscure. Much of genetics depends upon the operation of chance, as you have seen, and the rules of probability take on great importance. In this section, we group relevant rules of probability and a way of testing actual data for their conformity with genetic hypotheses.

1. Multiplication rule

Our knowledge of chromosome behavior allows us to predict the proportions of progeny of different classes. Segregation yields roughly equal numbers of each allelic class. In simple arithmetic, we have 1/2 or 0.5 of each class. When we predicted the outcome of crosses involving genes that assorted independently, we took advantage of what is called the multiplication rule. The multiplication rule tells us that **the probability of independent events occurring together is the product of their separate probabilities.** The probability of both A (probability p) and B (probability q) is pq. A good thing to remember is that the multiplication rule is used in "both . . . and" cases. Thus if genes A/a and B/b are independently assorting, AB gametes (having both A and B) is $0.5 \times 0.5 = 0.25$.

2. Addition rule

Another rule prevails in "either . . . or" cases. The addition rule states that **the probability that either one or the other of two mutually exclusive events will occur equals the sum of their separate probabilities.** The probability of either allele A (probability p) or allele a (probability q) is $p + q$: $0.5 + 0.5 = 1.0$. **Caution:** we would misapply this rule if we were to use it to predict the probability of gametes carrying either allele A of one gene or allele B of another, because A and B are not mutually exclusive.

3. The binomial

The expression $(p + q)^2$ can be used to find the probabilities of all permutations and combinations of genes and their alleles above and gives some insight into the joint use of the multiplication and addition rules. In a cross of two monohybrids ($Aa \times Aa$), the frequencies of both parents' gametes are $0.5\ A$ (probability p) and $0.5\ a$ (probability q). If we expand the binomial (multiply $[p + q]$ eggs $\times$ $[p + q]$ sperm), we are using the multiplication rule: AA zygotes, for example, contain both an A allele from one parent and an A allele from the other. We are also using the addition rule; the Aa genotype forms in two ways: either A sperm and a egg **or** the reverse. The binomial, when expanded, is $p^2\ AA + 2pq\ Aa + q^2\ aa$ with the probabilities of p and q each equal to 0.5. Thus our genotype frequencies will be $0.25\ AA$, $0.5\ Aa$, and $0.25\ aa$.

To see this rule in another context, consider the probabilities of families with two children with various birth orders of boys (M: probability $p = 0.5$) and girls (F: probability $q = 0.5$). These probabilities are $(p + q)^2 = p^2$ MM (0.25) $+ pq$ MF (0.25) $+ qp$ FM (0.25) $+ q^2$ FF (0.25). If birth order is unimportant, the distribution is 0.25 MM, 0.5 (FM and MF), and 0.25 FF, adding FM to MF for families with children of different sex.

4. Combinations

The expression $(p + q)^n$ is the general form of the binomial, and it allows one to calculate by expansion all specific combinations, regardless of order, of the events that the probabilities p and q refer to. Thus for the combinations of male (M, frequency p) and female (F, frequency q) children in three-child families, we would expand the binomial $(p + q)^3$ to $p^3 + 3p^2q + 3pq^2 + q^3$, the terms describing families with three males, two males and one female, one male and two females, and three females, respectively.

While there are more sophisticated ways of obtaining the coefficients of the terms of the expansion, a handy representation of the coefficients is Pascal's triangle. Here, the powers (*n,* above) are arranged vertically, and the coefficients for the terms are derived by adding the two immediately above. In the expansion, the powers of *p* fall from *n* to 0, and those of *q* rise from 0 to *n*.

Power (*n*)	Coefficients	Expansion
0	1	1
1	1 1	$p + q$
2	1 2 1	$p^2 + 2pq + q^2$
3	1 3 3 1	$p^3 + 3\,p^2q + 3\,pq^2 + q^3$
4	1 4 6 4 1	$p^4 + 4\,p^3q + 6\,p^2q^2 + 4\,pq^3 + q^4$

5. The chi-squared (χ^2) test

In practice, the progeny ratios of crosses are the means by which hypotheses are tested or distinguished. Crosses may exclude one hypothesis or satisfy the predictions of another. However, no geneticist can expect crosses to yield "expected" values exactly. Deviations of different magnitudes from expected ratios, if they are wholly due to chance, can in fact be predicted on purely mathematical grounds. Geneticists need a test by which they can estimate the probability that the deviations from expected values are due to chance, and a rule of thumb to identify probabilities so low that they cast doubt upon the hypothesis tested in the cross. For these purposes they use the **chi-squared (χ^2) test.**

   a. Calculating χ^2.   The value of χ^2 is determined by first finding the differences between the **numbers** (not percentages or frequencies) **expected** in a cross and the numbers **observed** for each genetic category. Each difference is then squared and divided by the expected number. The sum of these quotients is χ^2. The deviations are conveniently rendered positive by squaring them. By dividing squared deviations by the expected number, the deviations in categories of greatly different size become comparable to one another.

   b. Determining the degrees of freedom and finding *P*.   The probability of achieving the same or greater χ^2 value (a composite measure of deviation) by chance alone is found as a *P* **(probability) value** in a table or chart of χ^2 (Fig. 5–2). These give χ^2 values on the horozontal and the corresponding *P* value on the vertical. The chart or table asks for another parameter, namely the degrees of freedom, which is related to the number of categories in the analysis. The **degrees of freedom** in these examples is usually the number of categories minus one. (Once the numbers of individuals in all but the last category is known, the size of the last one is fixed.) The χ^2 and *P* values are related by a family of curves, each representing a different degree of freedom.

   A high value of *P* is associated with a low value of χ^2. A high *P* value means that chance alone is likely to explain the deviations in the data from an ideal distribution. For instance, the *P* value 0.90 means that of all possible χ^2 values, 90% of them will be equal to or greater than the one found. When the *P* value is 0.05 (5%, or one in twenty), and certainly when it is as low as 0.01, it becomes much more likely that factors other than chance underlie the deviations. Geneticists generally take *P* values of 5% or greater to indicate that deviations are probably due to chance. Values of *P* below 5% are called significant and values of 1% or less are "very significant" and indicate that deviations are not likely to be due to chance alone. In this way, the experimenter calls attention to the possible or probable existence of additional, non-random factors that cause deviations from the expected numbers.

   c. Examples of the use of the χ^2 test.   Let us imagine that we tried to repeat one of Mendel's experiments by testcrossing the F_1 hybrids produced by the mating of smooth,

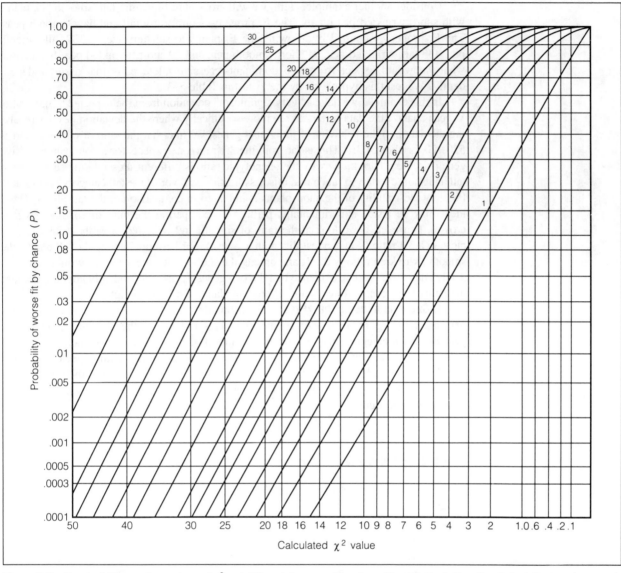

Figure 5–2 Graph of *P* for various values of χ^2 and different degrees of freedom. (From D. Hartl, Genetics, 3rd ed. Boston: Jones and Bartlett Publishers, 1993.)

yellow and wrinkled, green peas. Let us say the progeny were 132 smooth, yellow, 105 smooth, green, 111 wrinkled, yellow, and 96 wrinkled, green. It would be difficult to decide whether these data were consistent with a 1:1:1:1 ratio of the four phenotypes, or whether other factors, such as differential viability, had perhaps distorted the ratio. In Table 5–2, the operations of the χ^2 test are displayed, and the *P* value, drawn from the graph of χ^2 values on the line for 3 degrees of freedom (d.f.), is shown.

This table reveals that the deviations from the expected ratio can be considered insignificant; they could be due wholly to chance. The data indicate no need to investigate the matter further, and we can consider the data consistent with the expectations of independent assortment. However, **the hypothesis is not proven** by the test, as we will see in the discussion below.

Consider another example. Here, we will cross pure-breeding, tall, smooth pea plants with short, wrinkled pea plants. The F_1 progeny is uniformly tall and have smooth peas. The F_2 progeny, obtained by allowing the F_1 plants to self-fertilize, is 387 tall, smooth pea, 133 tall, wrinkled pea, 91 short, smooth pea, and 29 short, wrinkled pea. Is this ratio what we would expect from two dominant-recessive allele pairs that assort independently? We test this hypothesis by the χ^2 test in Table 5–3.

In this example, we have a "very significant" deviation from the expected 9:3:3:1 ratio. We can analyze the data further to see more specifically where the deviation lies. By resolving the data into the behavior of individual genes, we find that the ratio of tall and short is 520:120, or about 4.3:1. This is far from the 3:1 ratio one expects for a dominant-recessive gene pair. We do not know why the phenotypic ratio is different. Possibly the categories of tall and short overlap so much that misclassification in favor of the tall category is common. Possibly the viability of the short plants is impaired in comparison to the tall plants. These ideas may be tested experimentally, perhaps by comparing the outcome of homozygous crosses. However, if we look at the ratio of smooth and wrinkled, it is about 2.95:1, very close to a 3:1 ratio (you can test this with a χ^2 test). Furthermore, among the tall plants, the smooth : wrinkled ratio is 2.9:1, and among the short plants, the smooth : wrinkled ratio is 3.1:1. Thus there is some evidence that the two genes assort independently after all, because the ratio of pea-coat textures is not different in plants of different height.

d. Cautions in using the χ^2 test.   In testing data with the χ^2 test, you must use numbers not ratios, percentages, or frequencies. This is because the absolute numbers expected in each category play a large part in modifying the impact of deviations. Second, unless special corrections are used, one must exclude categories with expected numbers less than 5, either by excluding such categories entirely or by summing them with other categories. The test must be modified, particularly in the degrees of freedom, to take such changes into account.

Table 5–2 χ^2 test of data according to the hypothesis of independent assortment in a testcross

Phenotype	Number Observed (O)	Number Expected (E)	(O − E)	(O − E)^2	(O − E)^2/E
Smooth, yellow	132	111	21	441	3.97
Smooth, green	105	111	−6	36	0.32
Wrinkled, yellow	111	111	0	0	0.00
Wrinkled, green	96	111	−15	225	2.03
TOTAL	444	444			$\chi^2 = 6.32$
					d.f. = 3
					$P = 0.1 - 0.15$

Table 5–3 χ^2 test of data according to the hypothesis of independent assortment in the F_2

Phenotype	Number Observed (O)	Number Expected (E)	(O − E)	(O − E)^2	(O − E)^2/E
Tall, smooth	387	360	27	729	2.03
Tall, wrinkled	133	120	13	169	1.41
Short, smooth	91	120	−29	841	7.01
Short, wrinkled	29	40	−11	121	3.05
TOTAL	640	640			$\chi^2 = 13.48$
					d.f. = 3
					$P = {<}0.005$

A common mistake in the use of the χ^2 test is the assumption that an "acceptable" P value proves a hypothesis. It cannot do so; it simply tells the investigator that chance alone could account for the deviations from her predictions. To make this clear, consider a series of litters of two black rabbits, in which the ratio of progeny was 17 black and 7 white. Most geneticists would presume, for simplicity, that both parents were heterozygous (*Bb*), that the black offspring had genotypes *BB* or *Bb,* and that the white phenotype (*bb*) was recessive. If we test this "3:1" hypothesis with a χ^2 test, using a hypothetical 18:6 ratio, $P = 0.7$–0.8 with one degree of freedom. Suppose now that testcrosses of the black progeny in this case revealed that they were all heterozygous, and that further analysis revealed that the "dominant" homozygotes (*BB* in this example) die long before birth. The numbers of progeny (17 black and 7 white) should then be consistent with a ratio of 2 *Bb* to 1 *bb,* or 16 black : 8 white. A χ^2 test gives (with one degree of freedom) a P value between 0.7 and 0.8. This demonstrates that the test, even if it gives an acceptable P value for one hypothesis, does not exclude another interpretation of the same data.

By the same token, one cannot wholly discount a hypothesis that yields a low P value with a given set of data. As we saw in the example having to do with the distorted tall : short ratio in pea plants above, the deviations may be caused by factors having nothing to do with the idea being tested—or, in other cases where extreme deviations may occur, they may be due to chance after all.

REVIEW QUESTIONS AND PROBLEMS

1. In diploids, what is the genotypic ratio of the testcross progeny of a dihybrid (double heterozygote), in which the two genes are unlinked?

2. What is the genotypic ratio of the meiotic products of dihybrid cells (*AaBb*) in which the two genes assort independently?

3. A plant heterozygous for three unlinked recessive genes (genotype *AaBdDd*) is self-crossed. Give the proportions of the following genotypes or phenotypes: (i) *AaBbDd* (ii) *A_B_dd* (iii) *AAbbDD* (iv) *AABbdd.*

4. A cross of red yeast cells of mating type **a** to tan yeast cells of mating type α yielded tan diploids. When these diploid cells were induced to undergo meiosis, random meiotic products were: 103 red **a**; 99 tan **a**; 109 red α; and 100 tan α. What can you conclude from these results in terms of dominance of the color trait? Diagram the cross showing parents, zygote, Metaphase I, and meiotic products.

5. In our consideration of Mendel's results, we saw that homozygous smooth, yellow peas mated with wrinkled, green peas gave rise to F_1s that were smooth and yellow. The F_2 peas also included smooth yellow plants, as well as the other three possible combinations of characters. A self-cross of one of the smooth, yellow plants of the F_2 yielded 135 smooth, yellow and 47 smooth, green peas. What is the genotype of the plant that was self-crossed?

6. What is the basis of independent assortment?

7. The *A, B,* and *C* genes of corn assort independently. *A, B,* and *C* are completely dominant over their respective alleles *a, b,* and *c.* In the cross *AaBbcc* × *aaBbCc,* what proportion of the progeny are expected to be (i) *AabbCc;* (ii) *A_B_cc;* (iii) *aabbcc;* (iv) *A_B_C_?*

8. A cross of two corn plants, one tall with yellow kernels, the other short, with white kernels, yielded an F_1 that was tall and had yellow kernels. The intercross of the F_1 gave rise to an F_2 with the following phenotypes: 30 tall, yellow; 8 tall, white; 6 short, yellow; and

4 short, white. Do the determinants of these characters assort independently? Test the hypothesis with a χ^2 test.

9. A mating of two yeast strains, one having a red color (*red⁻*) and the other unable to use the sugar galactose as a carbon and energy source (*gal⁻*), yielded a progeny of 756 normal, 685 *red⁻*, 661 *gal⁻*, and 650 *red⁻ gal⁻*. Interpret this cross, and test your ideas by the χ^2 method.

10. In four-child families, what proportion of them have at least one boy?

11. In families with two girls and one boy, what fraction will have the boy as the second child?

12. In all families with four children, what fraction will have the gender order MFFM?

Chapter 6

GENES, PROTEINS, MUTATIONS, AND DOMINANCE

A gene has a sequence of nucleotides encoding the sequence of amino acids in a polypeptide, of which proteins are made. Three nucleotides in order unambiguously specify each of the twenty amino acids found in proteins. Most proteins are enzymes, the catalysts of almost all biochemical reactions. Other proteins become structural components of the cell or its membranes. Mutations often change or abolish the functions of proteins, and dominance relationships between alleles will depend heavily upon their relative degree of expression in heterozygotes. An allele that encodes a normal protein, active enough to mask the absence of activity of the mutant form of the protein, will be dominant. Alternatively, the normal and mutant alleles may be semi- or codominant, so that both have visible effects upon the phenotype of the heterozygote. Most genes will have mutant alleles of both types, at least at some level of the phenotype.

A. Proteins

1. Metabolic pathways

Many proteins contribute to the structure of the cell. Most proteins, however, are biological catalysts called **enzymes** that enable cells to perform specific chemical reactions. While proteins can catalyze reactions and contribute to cell structure, they cannot replicate. Conversely, DNA carries information and directs its own replication, but it is not a catalyst. These points highlight the chemical basis of the difference between the phenotype and the genotype.

The simplest cells grow at the expense of relatively few substances in the environment, from which they make a great variety of different compounds. They do so by combining and altering various molecules in many enzymatic reactions. These reactions form **metabolic pathways** which diversify a few starting materials into all of the small (low-molecular-weight) compounds needed by the cell for increases in mass, specialized products and energy metabolism (Fig. 6–1). A different enzyme catalyzes each chemical reaction of these pathways. There are more than a hundred pathways, each with many reactions, and a number of interconnections exist among the pathways.

The major metabolic pathways produce the small building blocks of macromolecules (Fig. 6–1). As we saw in Chapter 1, macromolecules fall into four main classes:
  i. **nucleic acids:** long, unbranched chains of ribo- or deoxyribonucleotides (four kinds in each nucleic acid);
  ii. **proteins:** having one or more long, unbranched **polypeptide** chains made of **amino acids** (twenty different kinds);

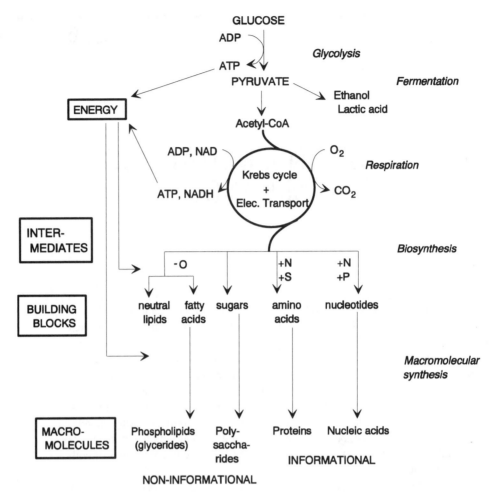

Figure 6–1 Summary of major metabolic activities in cells. Many steps shown are reversible by some route. In particular, macromolecular degradation is a common source of building blocks in reshaping cells and organisms.

iii. **polysaccharides:** long branched or unbranched chains of various sugars; and

iv. **lipids:** highly reduced carbon compounds, among them esters of fatty acids.

Nucleic acids and proteins are informational macromolecules. They carry information in specific sequences of nucleotides or amino acids, respectively. **The nucleotide sequences of DNA specify the amino acid sequences of proteins.**

2. Protein structure

Each of the twenty amino acids has an amino group and a carboxyl group attached to an **alpha carbon** (Fig. 6–2). To the alpha carbon is also attached one of twenty different chemical groups (**R groups**). The R groups are chemically diverse, and therefore the long sequences of these amino acids, the polypeptides, are even more diverse.

The four main aspects of protein structure are listed in Table 6–1. **The sequence of amino acids, the primary structure, is the only attribute of protein structure directly encoded by genes.** The other aspects of structure, having to do with the shape of the polypeptide and aggregation of polypeptides, are properties of the polypeptide chains themselves once they are made.

In a polypeptide chain, the basic amino group of one amino acid is attached to the acidic carboxyl group of the next. The bond is called a **peptide bond** (Fig. 6–2). Polypep-

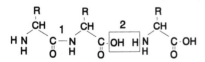

Figure 6–2 The peptide bond between amino acids (1), formed by removing water from the groups involved (2). The alpha carbons of amino acids are the ones to which the amino, carboxyl, and R groups are attached.

Table 6–1   Aspects of the structure of a protein

Level of structure	Definition	Origin
Primary	Amino acid sequence	Nucleotide sequence of the corresponding gene
Secondary	Polypeptide coiling or sheet formation	Hydrogen bonding between elements of the polypeptide backbone
Tertiary	Three-dimensional shape of polypeptide	Chemical interaction among the R groups of the amino acids in the polypeptide
Quaternary	Number and types of polypeptides (subunits) in the protein molecule	Aggregation of folded polypeptides at complementary surfaces. Polypeptides usually adopt symmetrical arrangement

tide chains may be as short as twenty amino acids (below which they are often called oligopeptides or peptides) or as long as 2000 amino acids. The sequence of amino acids in a polypeptide is its **primary structure.** The polypeptide adopts its **secondary structure** according to natural interactions of the atoms of the peptide-bond backbones, leading to local coiling or sheet formation. The overall shape of polypeptides, the **tertiary structure,** depends upon interactions of R groups in different parts of the chain, stabilized largely by weak bonds such as hydrogen bonds, hydrophobic (water-excluding) bonds, and salt linkages. One strong, covalent **disulfide** bond is found in the tertiary structure of many proteins. The tertiary structure underlies all protein interaction and catalytic function. The exact shape of the molecule provides a surface that recognizes small or large molecules. Enzymes have one or a few small regions called **active sites,** which are the points at which the enzyme catalyzes very specific biochemical reactions.

Finally, the **quaternary structure** is the **subunit** composition of the protein. Most proteins are made up of more than one polypeptide chain or subunit. Most multichain proteins have chains of the same type, arranged as symmetrical dimers, trimers, tetramers, etc. Others, quite numerous, have two or more kinds of polypeptide. Hemoglobin is one such protein, having two α and two β polypeptides. The interactions among R groups responsible for tertiary structure also stabilize the quaternary structure.

Generally speaking, each gene encodes a single polypeptide. This generalization emerged from research showing that mutations caused defects or the abolition of a single enzyme activity among large numbers of reactions in the cell. At that time, the "one gene-one enzyme" rule was the form this principle took. However, many proteins like hemoglobin have more than one polypeptide, and in all cases, distinct genes encode each one. Therefore, the principle is now embodied in the phrase, **"one gene-one polypeptide."**

B. The Genetic Code

1. Nucleotides and amino acids

What is the relation between the four-nucleotide code of DNA and the twenty-amino acid language of polypeptides? The genetic code is the "dictionary" governing how groups of nucleotides unambiguously specify individual amino acids in polypeptides. The actual biochemical steps of protein synthesis are very complex, and we explore them in Chapter 15. For our present purposes, it is sufficient to know that gene expression consists of two steps. The first is called **transcription,** in which the sequence of nucleotides of one DNA strand of a gene is copied into a complementary sequence of nucleotides of an RNA molecule (Fig. 6–3). Such RNA molecules are called messenger RNA, or mRNA. RNA is single-stranded and copied from only one of the two chains of DNA by the base-pairing mechanisms we saw in replication. RNA has ribose rather than deoxyribose as its sugar, and in place of thymine (T), it has a very similar base called uracil (U). However, it resembles a single

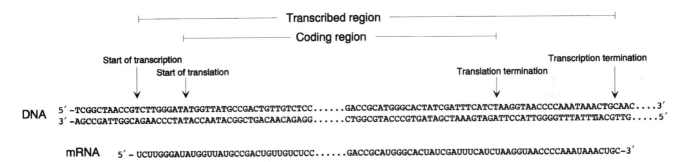

Figure 6–3   The coding and polarity relationships between a gene (DNA), its transcript (mRNA) and its translation product (protein). The sequences have been compressed. In fact, the untranslated regions before and after the coding region are normally considerably longer than shown, particularly in eukaryotes.

chain of DNA in its general structure, and its nucleotide bases (A, G, C, and U) have similar base-pairing properties as DNA nucleotides (A, G, C, and T).

The second step of gene expression is called **translation** because the four-nucleotide code of mRNA serves as the source of information in protein synthesis. This step is a real translation, since a four-nucleotide language of mRNA is transformed into the twenty-amino acid language of polypeptides.

2. Formalities of translation

Each amino acid requires at least three nucleotides in sequence to specify it unambiguously. A 1-to-1 relationship could encode only four amino acids, one for each nucleotide. A 2-to-1 relationship would also be insufficient, because there are only sixteen different dinucleotides (4 in the first position × 4 for each in the second position). There are, however, 64 (4 × 4 × 4) different trinucleotides, or **triplets,** which is more than enough. The genetic code, illustrated in Figure 6–4, is in fact a triplet code. Another term for a triplet in this context is **codon,** meaning a unit of the genetic code.

In translation, mRNA is "read" arbitrarily, starting with a codon for the amino acid methionine, AUG (*see* Figs. 6–3 and 6–4). Prokaryotes and eukaryotes identify the appropriate AUG as a start somewhat differently, but in doing so, they define the proper **reading frame.** In a triplet code, there are three reading frames depending on which nucleotide is chosen as a starting point. The choice of an AUG as a start arbitrarily defines the reading frame.

With 64 triplets for twenty amino acids, there appear to be 44 more than are strictly needed. Three of these triplets (UAG, UGA, and UAA) are noncoding. These codons are called nonsense codons. They cause the polypeptide chain to be terminated, bringing the polypeptide to completion. The other "excess" triplets are synonymous codons for various amino acids. As Fig. 6–4 shows, some amino acids are encoded by 6 different triplets, others by 4, 3, 2, or 1.

C. Mutations

To understand gene expression, it helps to understand the effect of mutations in a gene upon the protein specified by that gene. Mutations often arise as errors in DNA synthesis. Mutations affecting one or a few nucleotide pairs fall into two general categories: **base-pair substitutions** and **frameshifts.** They are called **point mutations** because of their localized nature. A diverse third category, macrolesions, comprise severe alterations such as translocations, inversions, deletions, insertions or alterations in chromosome number. These will be described in Chapter 10. Table 6–2 describes the three general categories of mutation.

mRNA nucleotide triplets, 5′ to 3′

Second nucleotide

First nucleotide	U	C	A	G	Third nucleotide
U	Phe	Ser	Tyr	Cys	U
	Phe	Ser	Tyr	Cys	C
	Leu	Ser	Non	Non	A
	Leu	Ser	Non	Trp	G
C	Leu	Pro	His	Arg	U
	Leu	Pro	His	Arg	C
	Leu	Pro	Gln	Arg	A
	Leu	Pro	Gln	Arg	G
A	Ile	Thr	Asn	Ser	U
	Ile	Thr	Asn	Ser	C
	Ile	Thr	Lys	Arg	A
	Met	Thr	Lys	Arg	G
G	Val	Ala	Asp	Gly	U
	Val	Ala	Asp	Gly	C
	Val	Ala	Glu	Gly	A
	Val	Ala	Glu	Gly	G

Figure 6–4 The genetic code. The nucleotides of the triplets are read in the order indicated on the sides of the chart. For instance, AUG is the codon for methionine (Met).

Table 6–2 Types of mutation

Type of mutation	Subclass	Genetic alteration	Effect(s)
Base substitution	Transition	Purine to purine and pyrimidine to pyrimidine	Mutation to nonsense codon (chain termination);
	Transversion	Purine to pyrimidine or reverse	Missense codon (possible defective protein); Synonymous codon (no effect)
Frameshift	Addition	Changes length of DNA and mRNA by a few (one to four) nucleotides	Out of frame: ribosome fails to read normal code for protein
	Deletion		In frame (+3 or −3 nucleotides): addition or deletion of an amino acid, with possible changes at the site(s) of mutation
Macrolesions	Change in chromosome structure	Inversions, translocations, deletions, duplications (see Chapter 10)	
	Change in chromosome number	Aneuploid and euploid variations (see Chapter 10)	

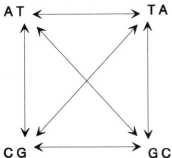

Figure 6–5 Base-substitution mutations. The diagonals of the box represent transitions; the sides represent transversions.

1. Base-pair substitutions

Mutations of this type are substitutions of one nucleotide by one of the other three nucleotides (Fig. 6–5). The substitution of a purine for a purine or a pyrimidine for a pyrimidine is called a **transition** mutation (*see* diagonals in Fig. 6–5); the substitution of a pyrimidine for a purine or a purine for a pyrimidine is a **transversion** (*see* sides in Fig. 6–5). The figure makes clear that an "A-T" base-pair is not equivalent to a "T-A" base pair in the context of a DNA double helix.

Base-pair substitutions will change amino acid codons in protein-coding regions of genes. The effects of such base-pair substitutions vary greatly. At the least consequential, one codon is altered to a synonymous codon. For instance, the change from the codon GGG to GGA by a transition in the third position does not change the amino acid in the corresponding protein; both codons encode glycine (Fig. 6–4). Even if an amino acid change occurs, it may be tolerated because it is a "conservative" substitution (the hydrophobic leucine may be replaced by the hydrophobic isoleucine), or because the substitution occurs in a part of the protein that is not important to folding, stability, or function.

Amino acid substitutions in the active site of an enzyme, or those in regions imparting stability to the protein, have more drastic effects. Even more serious are base-pair substitutions that change meaningful ("sense") codons into one of the three nonsense or chain-terminating codons: UAG, UAA or UGA. The effect of these mutations is severe if they occur anywhere in a protein-coding region except the most C-terminal part. The protein produced lacks all amino acids after the position of the nonsense codon. Truncated proteins arising in this way are usually quite unstable and disappear soon after they are synthesized. The effects of various base-substitution mutations are illustrated in Figure 6–6.

2. Frameshift mutations

Frameshifts, the other general category of point mutation, are additions or deletions of 1 to 4 consecutive nucleotide pairs in a DNA molecule. The effect of frameshifts is severe, because frameshifts of 1, 2, or 4 nucleotides change the reading frame. In protein synthesis, mRNA molecules are read arbitrarily from a fixed starting point, the first AUG codon (methionine) of a mRNA (*see* Chapter 15). The next three nucleotides are read as the second codon, and the next three the third. Removal or addition of one or two nucleotides will shift the reading frame such that the genetic message originally carried by the mRNA after the point of mutation is not read at all. Figure 6–6 illustrates the effect of frameshift mutations on the corresponding protein.

There are two abnormal reading frames (+1 and +2), achieved by adding one or two nucleotide pairs to a gene. After the point of mutation, the mRNA is read incorrectly, and reading usually stops before the end of the coding region because a nonsense codon is encountered in the abnormal reading frame. Occasionally, however, +3 or −3 frameshift mutations occur. These simply lead to net addition or deletion of one amino acid in a protein which may have a severe or mild effect, depending upon where in the protein they might occur.

3. The value of mutations

Only a minority of mutations improve the function of proteins. It is a minority because most beneficial changes have already occurred and remain highly selected in well-evolved systems. This means that organisms that suffer random changes in the code for an important protein are less, rather than more, likely to benefit from it. Nevertheless, improvements in a complex system are always possible. This is often the case if we recognize that organisms or their habitats may change such that mutations deleterious or inconsequential at one time might be beneficial at another. A spectacular example of a mutation that is both "good" and "bad" is one in the gene for the β chain of hemoglobin. The Hb^S allele, which impairs the oxygen-exchange properties of the molecule, also imparts to heterozygotes (Hb^A/Hb^S) a significant resistance to malaria. The Hb^S allele is found in many humans living in equatorial regions where the disease is prevalent.

The example of mutant hemoglobin is one of countless examples of a mutant, allelic form of a protein. The examples we have used in previous chapters, such as the *bw* muta-

```
WILD TYPE     ACC CAC UCU GGA UUG AAG GCA
              thr  his  ser  gly  leu  lys  ala
                       *
SAMESENSE     ACC CAU UCU GGA UUG AAG GCA
              thr  his  ser  gly  leu  lys  ala
                           *
MISSENSE      ACC CAC UCU GUA UUG AAG GCA
              thr  his  ser  val  leu  lys  ala
                           *
NONSENSE      ACC CAC UCU UGA UUG AAG GCA
              thr  his  ser  stop
                   *
FRAMESHIFT    ACC ACU CUG GAU UGA AGG CA
   (-1)       thr  thr  leu  asp  stop
                   *
FRAMESHIFT    ACC CAA CUC UGG AUU GAA GGC A
   (+1)       thr  glu  leu  trp  ile  glu  gly
```

Figure 6–6   Mutations (asterisks) in a DNA sequence and their effects (italicized) upon the corresponding amino acid sequence.

tion of *Drosophila* or the *red* mutation of yeast, arise in the same way. They have the effect of blocking a biochemical process: they lead to the impairment of an enzyme required for the synthesis of a pigment or an important metabolite. We must now put this knowledge into the context of dominance, in which mutant and normal alleles of a gene coexist in a single cell.

D. Complete Dominance

The word **dominance** applies only to **alleles of the same gene.** (Another word—epistasis—refers to the action of one gene upon the expression of another, non-allelic gene.) Accordingly, dominance can be studied only in diploids or in cells that contain two different copies of a given gene. A recessive allele usually reflects a loss of function with the dominant allele conferring a normal phenotype when present in a single dose, as in a heterozygote. This occurs because many genes make enough of the protein they encode, or the lower amount of enzyme makes enough of its product, to mask the inactivity of the protein encoded by a mutant allele. Our common notation for phenotypes, in which *A_* denotes the dominant phenotype and *aa* signifies the recessive, reflects this fact. **The criterion of complete dominance is that the homozygous dominant genotype and the heterozygote cannot be distinguished phenotypically.** The dominance of the *bw*+ allele over the recessive *bw* allele is an example of this relationship.

E. Incomplete or Semidominance

An incompletely dominant allele does not wholly mask the presence of another allele of the same gene. In cases in which the mutation leads to complete inactivation of the protein, partial dominance reflects the fact that the active allele does not have sufficient "excess capacity" to impart a normal phenotype to a heterozygote. In snapdragons, true-breeding red-(*RR*) and white-flowered (*rr*) plants produce pink F$_1$ hybrids (*Rr*). Here, a gene required in flower pigment formation is not quantitatively sufficient in one dose to give full red color to the flowers. An appearance intermediate between the parental phenotypes results.

F. Codominance

Codominant alleles are *qualitatively* distinct; each can be specifically detected in the presence of the other. Usually the word is applied to allelic forms of a protein. Human blood antigens are examples. For instance, the M and N antigens of blood cells have somewhat different amino acid sequences and can be distinguished by specific antibodies. With the appropriate antibodies, one can detect the presence or absence of either antigen. The three phenotypes of humans for the M/N series are M (genotype *MM*), N (*NN*), and MN (*MN*). The last genotype reacts positively to both specific antibodies, and the alleles are thus codominant. In contrast to a quantitative blend on a continuous scale, as in the example of snapdragon colors, the *M* and *N* alleles express themselves as discrete characters in heterozygotes.

G. Problems in Determining Dominance Relationships

1. Multiple alleles

It should not be surprising, given the large number of possible mutations a gene might suffer, that most well-studied genes have multiple alleles with a variety of effects on function. For most genes, the "normal" (wild-type) allele is dominant to most mutant alleles isolated in the laboratory. Many of these mutant alleles, however, do retain some of the normal function. While they remain recessive to the normal allele, they are incompletely dominant to complete loss-of-function alleles. Because they do not have the excess capacity of the normal, dominant allele, these partial mutations give a more normal phenotype as homozygotes (double dose) than as heterozygotes with a loss-of-function allele (single dose).

Mutants wholly lacking function are called null alleles or "amorphs." Null alleles include those in which part or all of the gene is actually deleted, as well as most nonsense mutations. Those retaining some normal function—usually having an amino acid substitution—we call partially active or "hypomorphs." Some mutations can derange the quantitative control of gene activity so that they are more active than normal. Among these are mutants that might be dominant to the normal allele.

A great diversity of alleles prevails for most genes in natural populations. The alleles found in natural populations differ from one another in inconsequential or very subtle ways. One would not expect mutations expressed in natural populations, subject as they are to natural selection, to have radical effects. Because of their usually subtle differences, most naturally occurring alleles are semidominant to one another at the level of the gross phenotype; they are also codominant to one another if one could distinguish the polypeptides they encode.

The diversity of alleles in nature arises through neutral mutation (mutations having insignificant effects on the overall phenotype) or through selected mutations that are advantageous only in some seasons, conditions, or habitats. The hidden genetic variability represented by diverse alleles in populations is the raw material of evolution, because variants that are inconsequential in one situation may impart much more significant differences in fitness in other conditions or genomes other than the ones in which they arose.

2. Dominance and codominance among multiple alleles

Some genes, notably certain blood antigen genes, have both codominant and recessive alleles. In the *ABO* blood-type series, alleles I^A and I^B, codominant to one another, determine three of four blood types, A, B, and AB. However, both I^A and I^B are dominant to the recessive I^O allele which yields no active antigen: the I^AI^A and I^AI^O genotypes are phenotypically the same in standard tests, and only I^OI^O genotypes can display the O blood type.

3. Dominance at different phenotypic levels

Dominance relations may differ at different levels of the phenotype. Some mutant alleles, as we have noted above, are recessive because the amount of enzyme produced by the normal allele is enough to impart the normal phenotype to a heterozygote. However, if the actual amount of enzyme activity (as opposed to the product of the enzyme reaction) is measured, only half the amount of activity might be seen. Therefore, at the level of enzyme activity, the mutant and normal alleles are incompletely dominant to one another. Taking this one step further, let us say the mutant that lacks activity nevertheless produces an abnormal, inactive form of the enzyme protein. One can often detect or visualize inactive variant forms of polypeptides with antibodies. Using antibodies, one often finds that the variants have an altered mobility in an electric field, which reflects substitution of amino acids with charged R groups. (The technique used to separate proteins in this way is called **electrophoresis.**) Such tests often reveal both the normal and the abnormal mutant protein in heterozygotes. At this level, therefore, the alleles would be codominant.

H. Dominance in Haploid Organisms

The assessment of dominance depends upon the phenotype of heterozygotes. It would seem impossible to do this in haploid cells, which cannot strictly speaking be heterozygous. Several tricks make it possible to assess dominance in haploids, even bacteria. All these tricks depend upon constructing cells that are partially diploid, or cells that contain haploid nuclei of different types. In these cases, simply having two alleles in the same cell—whether or not they are in the same nucleus—is all that is important. Once in the same cell, the two alleles can potentially contribute mRNA to a common cytoplasm, and the dominance relations of the alleles can be determined.

In bacteria, plasmids may carry an extra copy of a gene also found on the large circular chromosome. If the two copies of the gene are different, the dominance relations of the two alleles can be ascertained. The major problem here is controlling the number of copies of the plasmid in the cell. If a partially active allele were present in a multi-copy plasmid, it would appear more active phenotypically than if it were on the chromosome, only one copy of which is present. Some plasmids are present at 1–2 copies per cell, however, and these are widely used to make partial diploids to study dominance relationships in bacteria, as we will see in Chapter 19.

In filamentous fungi, the cells are actually tubes of cytoplasm with many nuclei in a common cytoplasm, a condition we call **coenocytic.** Within a species, genetically different cells can often fuse with one another to form **heterokaryons,** in which haploid nuclei nevertheless remain separate. Dominance relations between alleles of a gene can be judged in heterokaryons made in this way (*see* Chapter 17).

In many other well-studied haploid organisms, chromosomal aberrations may create partial duplications of certain chromosome parts which are often not harmful if they are small (*see* Chapter 10). If different alleles of a gene lie on the two copies of the duplicated segment, dominance relations can be assessed by the phenotype of the aberrant organism.

I. Other Terms Relating to Gene Expression

Epistasis denotes the effect of one gene on the expression of another gene. Mutations that prevent color formation, for instance, are epistatic to mutations that alter the phenotype from one color to another. Epistatic interactions, unlike dominance relationships, can be studied easily in haploid organisms. Interactions of duplicate genes and complementary genes, which we shall see in Chapter 7, are examples of epistasis.

Expressivity denotes the degree of expression of a gene or mutation. Frequently gene expression varies, even in individual organisms of similar genotype. In some cases, mutations do not gain any expression at all in a fraction of the population, even though the members of the population have the same genotype. This indicates that in some individuals, the activity of a gene does not reach a critical threshold required for phenotypic expression. The word **penetrance** is used to describe the fraction of individuals, all carrying the gene in question, that express the gene to any degree.

Pleiotropy denotes multiple effects of a gene. It is not surprising that genes affecting a fundamental process thereby affect many other processes in a complex organism. For instance, the *phenylketonuria* (PKU) mutation of human beings blocks the pathway for disposal of the amino acid phenylalanine. If this is not recognized at birth, and if a low-phenylalanine diet is not provided to the infant, the resulting accumulation of the phenylalanine derivative, phenylketone, leads to severe mental retardation and metabolic disorders. Hence a single mutation may affect many traits. Many complex phenotypes can be traced in this way to a change in a single, pleiotropic gene.

REVIEW QUESTIONS AND PROBLEMS

1. Assume that the genetic code is a triplet code but that codons differ in meaning only according to whether there is a purine or a pyrimidine in the last position. What is the maximum number of different amino acids such a code could specify?

2. How would you characterize proteins structurally?

3. What do enzymes do? To what class of macromolecule do they belong, and what other members of this class are there?

4. What is synonymy in the genetic code?

5. How many possible tetrapeptides (peptides containing four amino acids) are there?

6. How many different triplets would there be if DNA had six, instead of four bases?

7. How does RNA differ from DNA?

8. How many other codons can mutate, with a *single* base change, to become UGG?

9. By use of the code table, determine how many amino acid substitutions can arise owing to single-base mutations in the codon UUG.

10. Consider a DNA molecule with the following sequence of base-pairs in a protein-coding region.

    5′ AGGCTTACTGGCCT 3′
    3′ TCCGAATGACCGGA 5′

    What will the mRNA sequence corresponding to this sequence be, assuming that transcription starts from the left?

11. Mutations of single base pairs in DNA can abolish the ability of a gene to be expressed as the protein for which it is a code. Show two ways in which this might happen.

12. A friend wishes to know why base-substitution mutations in a gene vary so greatly in their effects on the corresponding protein. You have just a moment, and want to answer by giving him (i) an example of a wholly innocuous (not at all harmful) base-substitution, and (ii) an example of a base-substitution that renders the cell unable to make the protein.

13. A man with blood group M and a woman of blood type MN had three children: one MN, one M, and one N. Which child or children could the man reasonably suspect was not his own on the basis of blood groups?

14.   A man with the M blood type married a woman of blood type MN. What possible blood types could their legitimate children have?

15.   Distinguish dominance from semidominance by giving the experimental (practical or observable) criterion for each one.

16.   What is the basis for the fact that most laboratory mutations are recessive?

Chapter 7

MODIFIED GENETIC RATIOS: GENE INTERACTION

Dominance is a functional relationship between alleles. However, different genes will necessarily interact if they affect the same trait. Interaction of non-allelic genes is called epistasis. Epistatic relationships, together with semidominance, may greatly modify phenotypic ratios in simple crosses.

A. One Dominant, One Semidominant (Diploids Only)

Phenotypic ratios in multigene crosses may be hard to interpret if the genes segregating in the cross interact or if dominance relationships vary. In crosses involving two independently assorting genes, characteristic departures from the 1:1:1:1 phenotypic ratio of haploids, or diploids in testcrosses, or from the 9:3:3:1 F_2 ratio of diploids are seen for different gene interactions.

If one of two genes assorting in a dihybrid F_2 progeny is semidominant, the phenotypic ratio will be more complex than the ones we dealt with in Chapter 5. For instance, if the dominant-recessive pair of alleles, *A/a,* with phenotypes tall (*A_*) and short (*aa*), and the semidominant *B/b* pair of alleles, with red (*BB*), pink (*Bb*) and white (*bb*) phenotypes, were assorting independently, both the tall and the short plants would be distributed 1/4 red : 1/2 pink : 1/4 white. Thus the F_2 has the complex ratio:

3/16 tall, red	1/16 short, red
3/8 tall, pink	1/8 short, pink
3/16 tall, white	1/16 short, white

Upon encountering this ratio with no prior knowledge of the underlying genes, you must **resolve the progeny into classes, one character at a time,** and see how each character behaves. Use this familiar checklist: (i) What are the characters and the allelic forms? (ii) What is the ratio of phenotypes for each character? (iii) Do the genes assort independently or not? Let us answer these questions:

- The progeny display two heights and three colors.

- For height, the plants distribute themselves 3 tall : 1 short, suggesting one gene with complete dominance. For color, the ratio is 1 red : 2 pink : 1 white, suggesting a single pair of semidominant alleles for red and white.

• Independent assortment clearly prevails, because among both tall and short plants, we see a 1:2:1 ratio for red, pink, and white. Approaching it differently, we could see that among all three color classes, there is a 3:1 ratio of tall to short plants.

B. Complementary Functions

Most characters require the action of a number of genes. Biochemical pathways, which are sequences of enzyme reactions, are excellent examples of this principle. Mutations affecting complex functions reveal a great deal about the genetic organization and physiological roles of the corresponding enzymes, as we will see in Chapter 17. At this point, we need only remember that the lack of an enzyme at any point can block the formation of the end product.

1. Haploid example

Consider the synthesis of an amino acid by *Neurospora crassa,* a haploid fungus. Amino acids, such as arginine, are made by sequences of enzymes. If a mutation in a particular gene eliminates a particular enzyme of the pathway, the fungus grows only when it is given arginine in the culture medium. Now consider a mutation in a second gene, lying on a different chromosome from the first gene, that controls another enzyme of the arginine pathway. Strains carrying the second mutation also require arginine in order to grow.

When two mutant strains A and B, carrying mutations *arg-1* and *arg-2,* respectively, mate, the ratio of their progeny might puzzle you: the ratio is 1/4 normal (do not require arginine) to 3/4 that require arginine. This ratio is familiar, but do not be misled by it.

It is time for the checklist: (i) Are we dealing with a haploid or a diploid? You know that *N. crassa* is haploid. (ii) What are the characters and alleles controlling them? You have been given this information: the two mutations have occurred in different genes. Therefore each strain, although it carries a mutation for one of the genes, is normal for the other. Thus Strain A, which we might refer to casually as "*arg-1,*" has the genotype *arg-1 arg-2*$^+$ and Strain B ("*arg-2*") has the genotype *arg-1*$^+$ *arg-2.* Because the fungus is haploid, it can have only one allele of each gene. (iii) What phenotypic ratio can we expect in a cross of the two? Going through the previous steps of the checklist makes this much simpler because it gives you symbols to work with. Because the genes lie on different chromosomes, they will assort independently.

The genotypes of the parents and the meiotic products, given below, reveal that 3/4 of the meiotic products are unable to make arginine, owing to one or two mutations in the pathway, while 1/4 require no arginine: *arg-1 arg-2*$^+$ × *arg-1*$^+$ *arg-2* yields:

arg-1 arg-2$^+$	(requires arginine)
arg-1$^+$ *arg-2*	(requires arginine)
arg-1 arg-2	(requires arginine)
arg-1$^+$ *arg-2*$^+$	(requires NO arginine)

We will thoroughly explore a set of arginine mutations for a more sophisticated analysis of gene action in Chapter 17.

2. Diploid example

In the case of plant flower pigments, the normal flower color requires that all of the relevant biochemical reactions proceed properly. Consider the pathway below:

Enzyme 1 (gene *A*)   Enzyme 2 (gene *B*)

X ——————⟶ Y ——————⟶ blue pigment

A homozygous mutant for either gene *A* or gene *B* blocks pigment formation, just as mutations in the arginine pathway of *Neurospora* blocked arginine synthesis. Keep this mechanism of gene action in mind as you interpret the following crosses.

- Formation of the hybrid. Two homozygous white-flowered plants carrying recessive mutations were crossed. They gave rise to progeny with purple flowers. This information immediately tells you that the parents had different defects, and that both defects are recessive. If the mutations carried by the two parents were the same, the cross would yield a homozygous recessive phenotype, namely, white. At this point you must "translate" this information into symbols comparable to the haploid example of *arg-1* and *arg-2* above: the mutations might be called *wh-1* and *wh-2*. Follow this thought rigorously to determine that the diploid genotypes of parents (homozygous) and progeny are:

$$wh\text{-}1^+/wh\text{-}1^+ \ \ wh\text{-}2/wh\text{-}2 \ \times \ wh\text{-}1/wh\text{-}1 \ \ wh\text{-}2^+/wh\text{-}2^+$$

(white parent 1)   (white parent 2)

yields:

$$wh\text{-}1^+/wh\text{-}1 \ \ wh\text{-}2^+/wh\text{-}2$$

(blue F_1)

- The F_2. Upon intercrossing the blue F_1 progeny above, two phenotypes appeared in the F_2 in a peculiar ratio: 9 purple to 7 white. This ratio actually bears out the implications of the P and F_1 genotypes. Looking at the genotypes of the F_2, you will see—especially by listing them—that 9/16 of them carry at least one $wh\text{-}1^+$ and one $wh\text{-}2^+$ allele. The other 7/16 will be homozygous recessive for one or the other or both *wh* alleles. This is shown in the summary table (Table 7–1) at the end of this chapter.

- Method. Interpreting crosses, particularly when they have peculiar outcomes, is most efficient if genotypic symbols are hypothetically attached to phenotypes early in the process. In these two cases of complementary function, understanding the problem was an automatic outcome of translating simple information into precise symbolic form. In doing this, be careful to use symbols that allow you quickly to recognize allelic forms of the same gene and prevent confusion between alleles of different genes.

C. Duplicate Functions

Organisms occasionally perform a single biochemical function in several ways. Sometimes the gene for an enzyme is duplicated on another chromosome. In other cases an end product can be made from several different starting materials. Thus a function can be carried out by either of two non-allelic genes, often assorting independently. A mutation of one or the other cannot eliminate the function; two different mutations must be present simultaneously to do so.

1. Haploid example

In many strains of haploid yeast, there are two genes, on different chromosomes, that encode the enzyme maltase. Either gene in its wild-type form allows yeast to use the sugar maltose for growth. A strain with mutations in both of the maltase genes was isolated. A mating of this strain with a wild-type strain (having active alleles of both maltase genes) yields a diploid cell that can use maltose as a source of carbon and energy for growth. When this diploid strain undergoes meiosis and forms haploid progeny, 3/4 are phenotypically normal and 1/4 are unable to use maltose for growth. How do we interpret this, especially if we were lacking prior knowledge of the genetic basis of the trait?

The interpretation lies in recognizing that the 3:1 ratio of meiotic products violates the criterion, a 1:1 segregation, for a single gene. (Remember that meiotic products are the haploid phase of yeast.) If 1/4 of the progeny are unable to perform a function, you must consider that more than one gene must be segregating in the cross. The simplest case is two genes, assorting independently:

$$mal\text{-}1\ mal\text{-}2 \times mal\text{-}1^+\ mal\text{-}2^+$$

yields:

$mal\text{-}1^+\ mal\text{-}2^+$ (can use maltose)

$mal\text{-}1^+\ mal\text{-}2$ (can use maltose)

$mal\text{-}1\ mal\text{-}2^+$ (can use maltose)

$mal\text{-}1\ mal\text{-}2$ (cannot use maltose)

You should also realize that a cross of the two single mutants, both able to use maltose for growth, will give the same result as the mating of the double mutant and wild-type.

2. Diploid example

Many examples of duplicate function are known in plant pigment formation:

Enzyme 1 (gene A)

X ⟶ red pigment

X ⟶ red pigment

Enzyme 2 (gene B)

If the wild-type genes (A and B) show complete dominance over their mutant alleles (a and b), and if only one A or one B allele is needed for full color, then only the genotype *aabb* will fail to make red pigment and display a deficient phenotype. If a mutant (*aabb*) and a wild-type (*AABB*) are mated to form an F_1 (*AaBb*), the F_2 (the progeny of the cross *AaBb* × *AaBb*), will yield the ratio of 15/16 pigmented ($A\_B\_, A\_bb, aaB\_$) to 1/16 unpigmented (*aabb*) (*see* Table 7–1).

D. A More Common Case of Epistasis

Epistasis refers to the effect of one gene on the expression of another. Mutants of either gene controlling complementary functions are epistatic to the other, because a single mutation can block overall function, regardless of the allelic state of the other gene. Similarly, duplicate gene functions fit the definition: in the presence of A, one cannot determine whether the genotype at the other gene is $B\_$ or *bb*. Thus in both cases, the wild-type A allele is epistatic to the B gene, and vice-versa: the two genes are in fact mutually epistatic. A less symmetrical case of epistasis follows:

1. Haploid example

In *Neurospora*, the gene *cr-1* alters the appearance of the asexual spores from fluffy aerial masses to a flat, velvety surface on the surface of the colony. Another gene, *acon*, prevents the formation of asexual spores entirely. The *acon* mutation is epistatic to the *cr-1* gene because it precludes observation of spores.

2. Diploid example

In this example, the pathway leading to a normal red pigment in a plant includes a precursor with a different color:

$$
\begin{array}{ccc}
\text{Enzyme 1 (gene } A\text{)} & \text{Enzyme 2 (gene } B\text{)} & \\
X \longrightarrow & Y \longrightarrow & \text{Red pigment} \\
\text{(colorless)} & \text{(Yellow pigment)} &
\end{array}
$$

A plant with the *aa* genotype lacks color entirely, regardless of the state of the *B* gene. However, if the *A* gene is normal (in an *A_* plant), then the color will depend upon the allelic state of the *B* gene. If the *B* gene is also normal (*A_B_*), the plant will be red, but if it is mutant (*A_bb*), it will be yellow. If we perform a cross that yields a dihybrid (*AaBb*), and then self-cross it, the F_2 ratio will be 9 red : 3 yellow : 4 white (*see* Table 7–1). Because the *a* mutation blocks the expression of yellow pigment, and thus the difference between *B_* and *bb,* we say that mutation *a* is epistatic to the *B* gene.

E. An Idealized Case of Quantitative Inheritance

In the example of duplicate functions, we assumed the active alleles of two genes were dominant to their mutant alleles. Consider duplicate functions in which heterozygotes are exactly intermediate in phenotype, and that the two genes have quantitatively similar impacts upon the phenotype. We expect *Aabb* to be the same as *aaBb,* and *AAbb* to be the same as *AaBb.* In other words, the phenotype will depend only on how many "active" alleles (capitalized) are present.

Consider two genes controlling plant height: plants with the genotype *aabb* are 10 cm high and those with genotype *AABB* are 20 cm high. The F_1 plants (*AaBb*) in a mating of the pure-breeding 10- and 20-cm plants average 15 cm high. In the F_2, we expect progeny having from 0 (*aabb*) to 4 (*AABB*) active alleles, distributed among five different height classes between 10 and 20 cm. Their ratios (1:4:6:4:1, using the F_2 genotype list) are summarized graphically in Figure 7–1.

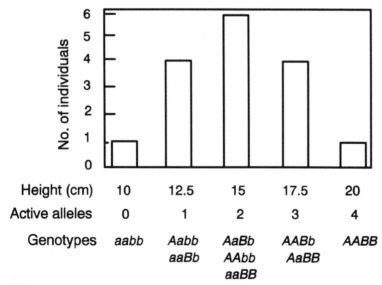

Figure 7–1   A simple case of quantitative inheritance for height.

Multiple genes underlie most metric traits (those measured on a continuous scale), and in most cases they have additive effects. Unlike this example, they need not have equal effects on the character. Moreover, their effects may not be confined to the trait in question: many different functions involved in growth and nutrition may also affect height. Finally, genes affecting a single character may be quite numerous, and ideal distributions like the F_2 above are rarely encountered, even in controlled experimental work, because other genes with minor effects may obscure differences caused by the two genes of interest.

If we add environmental variation to genetic variation, the problems of analyzing quantitative traits become even more difficult. Even among organisms of a single pure genotype there will always be a range of height, which will be narrower for populations grown in highly controlled growth conditions than for those in poorly controlled conditions. Thus genotypic classes will tend to overlap in phenotype. Only rarely can one resolve a two-gene distribution (with five genotypic classes in the above example) clearly by their phenotypes. Some of the methods of dealing with these matters are described in Chapter 13.

F. Summary of Modified Ratios

Table 7–1 gives most of the modified ratios discussed in this chapter, with the corresponding genotypic classes in each. In viewing this table, try to think of what each phenotypic group has in common at the level of the genotype.

Table 7–1 Phenotypes arising from gene interaction

Genotypic ratio	*A* and *B* dominant to mutants	*B/b* semi-dominant; *A* dominant	*A* and *B* complementary functions	*A* and *B* duplicate functions	*a* (mutant) epistatic to *B/b* gene
1 *AABB*	9 *A_B_*	3 *A_BB*	9 with *A* **and** *B*	15 with *A* **or** *B* or both	9 having *A* and *B*
2 *AaBB*					
2 *AABb*		6 *A_Bb*			
4 *AaBb*					
1 *AAbb*	3 *A_bb*	3 *A_bb*			3 having *A*, not *B*
2 *Aabb*					
1 *aaBB*	3 *aaB_*	1 *aaBB*	7 lacking *A* **or** *B* or both		4 lacking *A*
2 *aaBb*		2 *aaBb*			
1 *aabb*	1 *aabb*	1 *aabb*		1 lacking *A* **and** *B*	

REVIEW QUESTIONS AND PROBLEMS

1. Two true-breeding strains of sweet peas having white flowers were crossed. All F_1's had purple flowers. When the F_1's were self-crossed, the F_2 generation progeny were 875 with purple flowers and 725 with white flowers. Suggest a genetic explanation, and test it with a χ^2 test.

2. A white-flowered iris was crossed with a red-flowered iris. Both were true-breeding. The F_1 progeny were all red. The F_2 progeny were 145 red and 15 were white. Suggest a genetic explanation, and test it with a χ^2 test.

3. In wheat, a cross between red-kernel and white-kernel strains yielded F_1 offspring with red kernels. When the F_1 were intercrossed, the F_2 plants had a ratio of 15 red-kernel : 1 white-kernel. A testcross of the red-kernel F_1 plants yielded 3 red-kernel : 1 white-kernel. Explain.

4. Snapdragons with normal-shaped, red flowers are mated with plants with abnormal-shaped white flowers. In the F_1, all of the flowers have normal shape and are pink. Explain the outcome of the intercross of the F_1 plants that generated the following F_2:

   3/16 are red, normal

   6/16 are pink, normal

   3/16 are white, normal

   2/16 are pink, abnormal

   1/16 are red, abnormal

   1/16 are white, abnormal

5. A dominant allele, L, specifies short hair in guinea pigs and its recessive allele, l, specifies long hair. Codominant alleles at an independently assorting locus specify hair color, such that $C^Y C^Y$ = yellow, $C^Y C^W$ = cream, and $C^W C^W$ = white. In matings between dihybrid short, cream-colored pigs ($Ll\ C^Y C^W$), predict the phenotypic ratio expected in the progeny.

6. A true-breeding, short snapdragon with red flowers mated with a true-breeding, tall snapdragon with white flowers, yields F_1 progeny, all of which are medium height and pink-flowered. Assuming that only two pairs of semidominant alleles (for height and flower color) are assorting here, predict the proportion of F_2 plants that would have both medium height and pink-flowers.

7. *Neurospora* is a filamentous fungus normally having fluffy, orange masses of asexual spores called conidia. Two mutant strains, one having albino (white) conidia, and the other lacking conidia entirely (aconidial) were mated. Their progeny were as follows: 82 normal, 92 albino, and 166 aconidial. Explain these results as simply as you can, and test your explanation with a χ^2 test.

8. True-breeding red and blue irises were crossed and yielded a uniform crop of purple F_1 plants. These were intercrossed and the F_2 plants were 95 purple, 32 red, 35 blue, and 11 white. The white plants were crossed to the true-breeding, red parental strain and the progeny were all red. In the last cross, why were there no purple plants?

9. True-breeding white and purple irises were crossed; the F_1 progeny were all purple. When the F_1 plants were backcrossed to the white parents, the progeny were as follows: 24 red, 28 purple, 27 blue, and 25 white. What ratio of progeny would you expect if the F_1 plants were intercrossed?

10. True-breeding short irises with red flowers were mated to true-breeding tall irises with blue flowers. Their progeny were of medium height and had purple flowers. These F_1 plants were intercrossed and nine different kinds of plants appeared in the F_2. They were as follows:

41 medium, purple	20 medium, red	20 medium, blue
20 tall, purple	10 tall, red	11 tall, blue
19 short, purple	9 short, red	9 short, blue

If you were to back-cross the F_1 plants (medium, purple) to the short, red parent stock, what ratio of plants would you expect?

11. What is the difference between dominance and epistasis?

12. A yeast strain that required the amino acid arginine was mated to a normal strain, which could grow without arginine. The haploid meiotic products of the diploid, after meiosis, were 153 that required arginine and 15 that did not. Give an explanation for this peculiar result.

13. Certain recessive genes cause profound hereditary deafness, and individuals homozygous for such genes are occasionally found in high frequencies among extended families in small, isolated communities. The mutations originate in individuals several generations in the past, and become homozygous through marriages among relatives. A deaf man and a deaf woman from two different communities, each having deaf parents, had three children, all of whom had normal hearing. How would you explain this?

Chapter 8

SEX LINKAGE

Males and females of most higher animals differ in one of their chromosome pairs, the sex chromosomes. The female has two X chromosomes, and the male has an X and a functionally different Y chromosome. Genes on the X chromosome are called sex-linked. Sex-linked characters may be distributed differently in the two sexes of a progeny, and progeny of reciprocal crosses may differ.

A. Sex Chromosomes

Males and females of most higher animals are born in equal numbers. The 1:1 ratio, as in a testcross, suggests that one sex produces two types of gametes while the other sex produces only one. Because matings are confined to males × females, the 1:1 ratio persists generation after generation. In fact, this ratio reflects the existence of two kinds of **sex chromosomes.** Males of most animal species have different sex chromosomes, called X and Y, while females have two X chromosomes. Sex chromosomes are usually microscopically distinguishable from one another.

The existence of sex chromosomes called for a term for all the other chromosomes. Chromosomes that are not sex chromosomes are called **autosomes.** *Drosophila* has three types of autosomes and two sex chromosomes (Figure 8–1). The X and Y chromosomes behave like homologs in meiosis of males, pairing with one another in Prophase I and Metaphase I, even though they carry quite different genetic information. Therefore one-half the sperm will bear an X chromosome and one-half will bear a Y chromosome. Because all eggs produced by females contain an X chromosome, fertilization produces equal numbers of XX (female) and XY (male) zygotes. The word **heterogametic** refers to the sex with different sex chromosomes (males in *Drosophila* and humans). **Homogametic** is the term for the "non-segregating" sex (XX).

We see several departures from the *Drosophila* and human pattern in other organisms. Male grasshoppers are XO and females are XX. "XO" signifies that there is only one, unpaired X chromosome in diploid males, and at Anaphase I of meiosis, it goes to one or the other pole without pairing with any other chromosome. In fowl, many reptiles, and moths, females are heterogametic, having the sex-chromosome constitution WZ, and males are homogametic, ZZ. The letters Z and W are often used to call attention to the reversal of the usual hetero- and homogametic sexes. Sex-chromosome differences are even seen in some plants, although this is not common. There are many organisms that do not have a chromosomal sex-determining mechanism. Some of these organisms are hermaphroditic (having both sexes), and among such organisms, sexual compatibility may be determined by mating-type genes, as in yeast and *Neurospora*. Some organisms such as certain fish may have their sex determined environmentally during development or may even sustain an environmentally induced change of sex after they reach adulthood.

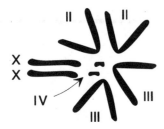

Female

Male

Figure 8–1 Sex chromosomes (X and Y) of *Drosophila*, seen in the rosette of chromosomes in the dorsal ganglion cells of the adult. The chromosomes of this unusual cell type are condensed in interphase, and the centromeres are gathered in the center of the nucleus (referred to as the chromocenter). The small fourth chromosome is among the smallest chromosomes known.

B. Sex-linked Inheritance

In organisms with the X-Y system, the X chromosome, like autosomes, carries genes that determine a large range of characters, many having no role in sex determination. However, the Y chromosome carries very few genes, and in *Drosophila* most are devoted to formation and maturation of sperm. The human Y chromosome carries a greater variety of genes, some with specialized functions and others actually homologous to X-borne genes. However, the Y chromosome nevertheless lacks many genes found on the X chromosome. As a consequence, genes on the X chromosome display a special pattern of inheritance called **sex linkage**. Because the male has only one copy of most X-borne genes, he is called **hemizygous** for those genes. Moreover, dominance of sex-linked genes cannot be observed in males, because two alleles of sex-linked genes cannot co-exist in normal males. The term sex linkage refers only to the pattern of inheritance; most genes on the X chromosome do not actually control sex-specific characters.

The pattern of inheritance of sex-linked genes differs from that of autosomal genes that we considered in previous chapters. It differs in two peculiar characteristics, which are not necessarily seen in every cross:

- The ratio of alternate characters may differ in the two sexes of the progeny.

- The progeny of reciprocal crosses may differ. **Reciprocal crosses** differ in the parent that contributes a given allele of a gene in a cross: *AA* females $\times$ *aa* males and *aa* females $\times$ *AA* males are reciprocal crosses. The sexes do not display different distributions of phenotypes in reciprocal crosses involving autosomal genes.

C. Crosses of *Drosophila* Involving a Sex-linked Gene

T. H. Morgan pioneered genetic investigations in the early 1900s, confirming Mendel's principles and extending them to other organisms as general principles. For this work, he won a Nobel Prize in 1933. Morgan's research team domesticated *Drosophila* for genetic studies, demonstrated clearly that genes were carried by chromosomes, discovered linkage and recombination of genes on the same chromosome, and fully explored the phenomenon of sex linkage. His work on sex linkage began when a white-eyed mutant he discovered in the normal red-eyed stock ("red-brown" of previous chapters) showed an anomalous pattern of inheritance.

1. The crosses

Morgan made reciprocal crosses of the white-eyed mutant, starting with true-breeding stocks:

- **Cross 1:** White females $\times$ red males yielded equal numbers of red females and white males in the F_1. **This result was not easily rationalized at first.** In the F_2, produced by intercrossing the F_1 males and females, a ratio of 1 red : 1 white appeared among both males and females. **This departed from the 3:1 F_2 ratio expected of an autosomal gene.**

- **Cross 2:** Red females $\times$ white males yielded all red F_1 progeny. In the F_2, males segregated 1 red : 1 white, while the females were all red-eyed. **This also differed from the F_2 progeny of crosses involving an autosomal gene,** despite the superficial 3 red : 1 white ratio in the progeny as a whole.

Notice that (i) phenotype distributions of the sexes differed in the F_1 progeny of Cross 1 and in the F_2 progeny of Cross 2; and (ii) the reciprocal crosses differed in their outcomes. As noted above, these are the criteria of sex-linkage.

2. Analysis of the crosses

Morgan inferred that white eye color was recessive because it failed to appear in the F_1 of Cross 2, but reappeared among the F_2s. With some of the reasoning below, he inferred that white eye color was due to a mutation on the X chromosome. Let us now make a symbolic representation of the crosses as we have in previous chapters:

- **Cross 1:** The white female parents of this cross must be diploid and homozygous, or w/w. The males in this cross must be hemizygous for red, or $w^+/\leftarrow$, where $\leftarrow$ signifies the Y chromosome. The F_1 of this cross is given in Figure 8–2. The figure shows that w eggs are fertilized by either w^+ or $\leftarrow$ sperm. The zygotes will develop into either heterozygous, red females (w^+/w) or hemizygous, white males ($w/\leftarrow$).

- **Cross 2:** The homozygous red female parents of Cross 2 must be w^+/w^+, and the white males must be $w/\leftarrow$. The F_1 of this cross is also given in Figure 8–2. Here, all zygotes receive a w^+ allele from their mother, and therefore the entire F_1 progeny is red.

The genotypes of the F_2 progeny of the two crosses were derived by randomly combining the two types of gametes of F_1 males and females in each case (Fig. 8–2). The genotypes derived in this way conform to the actual phenotypes of the F_2 progeny of the crosses, stated at the outset.

3. Some helpful points

Note three important points from the descriptions of the crosses above:

- Rigorous use of symbols, derived from simple observations of phenotype and knowledge of sex linkage, automatically rationalized the crosses.

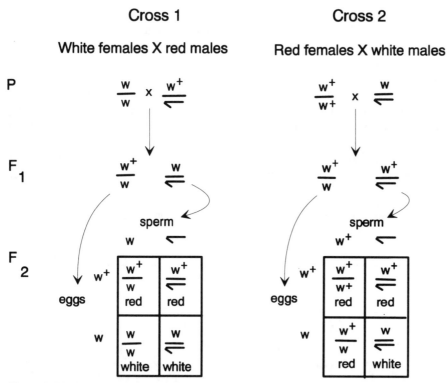

Figure 8–2 Reciprocal crosses of *Drosophila* involving the white-eye gene, through the F_2.

- Males always get their X chromosome from their mothers; males must get Y chromosomes from their fathers. Daughters, with two X chromosomes, must acquire an X chromosome from both parents.

- Recessive, sex-linked characters appear in males more frequently than in females. This is because recessive, sex-linked alleles are always expressed in the hemizygous male.

The two sets of crosses in Figure 8–2 illustrate all pertinent matings involving a sex-linked gene. Table 8–1 shows the same thing in simpler form. With this table, learn how to visualize quickly the progeny of any two parents.

To test your facility in resolving a problem into components, working with symbols and predicting the outcome of a cross, try to predict the results of a two-gene cross involving a sex-linked gene, white eye (*w*), and an autosomal recessive mutation, ebony body (*eb*), using the probability method: what are the F_1 and F_2 genotypes and phenotypes of true-breeding, ebony females mated to true-breeding white males?

D. Sex-linked Genes in Humans

Sex linkage in humans follows the rules outlined above for *Drosophila*. However, humans do not have large progenies, and therefore geneticists must infer the genetic basis of traits from **pedigrees**. Pedigrees, in the diagrammatic form shown in Figure 8–3, are often used to illustrate the inheritance of traits in family trees. The inheritance of sex-linked genes is particularly clear. For instance, if a phenotypic trait appears among the sons, but never among daughters of normal parents (Fig. 8–3), a sex linked, recessive mutation is almost certainly involved (Table 8–1, second line). If the trait does not appear among the children of the affected sons (assuming their wives are normal and homozygous), the point is confirmed (Fig. 8–3; Table 8–1, fourth line).

1. Hemophilia

Hemophilic people lack a protein needed for blood clotting. Minor cuts or internal bleeding are serious injuries for hemophiliacs, and many die from injuries that heal easily in normal people. Queen Victoria of England—who was not hemophilic—was a "carrier," or heterozygote, for this mutation. One of her sons suffered from the disease, but none of her daughters did. However, two of her daughters produced three hemophilic sons; four of her granddaughters were phenotypically normal, but produced among them, seven hemophilic sons. In fact, all hemophilia was confined throughout the pedigree to sons. The gene is easily traceable: all hemophilic sons implicate their mothers as carriers. The hemophilia allele can usually be traced back through the female line if fathers are all normal.

Hemophilia may disable or kill males before they can have children, although Victoria's hemophilic son Leopold survived to have a daughter. The daughter, Alice (normal

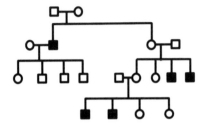

Sex-linked recessive

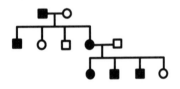

Autosomal dominant

Figure 8–3 Pedigree in which a sex-linked recessive is segregating (above) and, for comparison, a pedigree in which a dominant autosomal mutation is segregating (below). The conventions of representing pedigrees are shown: squares are male; circles are female; dark symbols represent the (mutant) phenotype in question. Some pedigrees do not depict the unrelated mates of the individuals of the family in question.

Table 8–1 All matings involving a sex-linked, recessive gene, and the phenotypic and genotypic outcome among male and female progeny

Parents	Daughters	Sons
$AA \times A{\leftarrow}^a$	All normal (*AA*)	All normal (*A*←)
$Aa \times A{\leftarrow}$	All normal (*AA*, *Aa*)	1 normal (*A*←) : 1 mutant (*a*←)
$aa \times A{\leftarrow}$	All normal (*Aa*)	All mutant (*a*←)
$AA \times a{\leftarrow}$	All normal (*Aa*)	All normal (*A*←)
$Aa \times a{\leftarrow}$	1 normal (*Aa*) : 1 mutant (*aa*)	1 normal (*A*←) : 1 mutant (*a*←)
$aa \times a{\leftarrow}$	All mutant (*aa*)	All mutant (*a*←)

^a The "hook" symbol (←) is used to represent the Y chromosome.

phenotype, but without doubt a carrier), had two hemophilic sons. Adult hemophilic women were rarely encountered in the past, in part because such women must have a hemophilic father and a carrier mother. Moreover, the menstruation attending puberty was a crisis for hemophilic women. (Menstruation has more recently been suppressed pharmacologically in hemophilic women in order to improve their survival.) Accordingly, few if any, hemophilic women left children.

Hemophilic people may now live more normal lives with the advent of methods for isolating the missing clotting factor from outdated blood from blood banks. The factor, when regularly injected, restores hemophiliacs' blood-clotting ability. Unfortunately, in the late 1970s and early 1980s, a considerable number of AIDS infections were transmitted by such injections, owing to the presence in the public blood supplies of the then unrecognized human immunodeficiency virus (HIV). The blood supply is now rigorously screened to test for HIV in order to prevent AIDS infections among hemophiliacs and the recipients of blood transfusions. There is now hope of making the clotting factor in abundance by recombinant DNA techniques, avoiding blood-derived supplies entirely.

2. Color blindness

This condition reflects the lack of one of the proteins required for color perception. Several recessive mutations on the X chromosome impart color blindness. Accordingly, it appears much more frequently in males than in females and obeys the rules of sex-linked inheritance. Unlike hemophilia, color blindness is a relatively benign condition.

Color-blind male children are born to heterozygous (carrier) or to homozygous, color-blind women. The father's genotype is irrelevant, because a son does not acquire his X chromosome from his father. Color-blind women can arise only among children of color-blind men and carrier or homozygous, color-blind mothers.

REVIEW QUESTIONS AND PROBLEMS

1. A color-blind man and his wife, with normal vision, have a color-blind daughter. What is the probability that their newborn son will also be color-blind?

2. The recessive gene *eb* (ebony body) is an autosomal mutation in *Drosophila*. The recessive gene *w* (white eyes) is sex-linked. True breeding ebony males are mated to true-breeding, white-eyed females. Predict the F_1 and the F_2 phenotypes and their ratios.

3. A sex-linked recessive gene, *let* (lethal), in *Drosophila* causes death during embryogenesis when *let* is homo- or hemizygous. If a female heterozygous for *let* is crossed to a wild-type male, what will the sex ratio of the adult progeny be?

4. A dairy farmer and a geneticist were looking out the farmer's living room window and saw a mahogany-colored Ayrshire cow with its newly born red calf. The farmer remarked that he was interested to learn the sex of the calf, and the geneticist told him that it was obvious from the color of the calf. She explained that in Ayrshires the genotype AA is mahogany and aa is red; but the genotype Aa is mahogany in males and red in females. What sex did she say the calf was?

5. Duchenne-type muscular dystrophy is an inherited human disease, and one type of color blindness is sex-linked. Two healthy parents with normal color vision have fully normal daughters. However, half their sons are healthy but are color-blind; the other half have normal color vision but have Duchenne's disease. Is Duchenne's disease an autosomal or a sex-linked trait? Diagram the probable genotype of the parents. Could such parents ever have a normal son? If so, how?

6.　In *Drosophila,* white (*w*) is a sex-linked recessive mutation that blocks incorporation of pigment into the eye. Scarlet (*st*) is an autosomal recessive mutation that blocks formation of brown eye pigments, thereby brightening the eye from the normal red-brown to a more crimson color. Predict the F_1 and F_2 phenotypes of the following matings, assuming that parents are true-breeding, single mutants: (i) white male × scarlet female; (ii) scarlet male × white female.

7.　You cross true-breeding roosters with no comb, with true-breeding hens with large combs. The difference in comb phenotype is caused by a single sex-linked trait with large-comb dominant. What will the phenotypes and their proportions be in the progeny of this mating?

8.　A homozygous, wild-type *Drosophila* female was mated to a homozygous male with cut wings and ebony body. Their progeny (the F_1) were normal females and males. The F_2s (the progeny of F_1 males and females) were as follows:

Females	Males
31 wild-type	16 wild-type
10 ebony	14 cut-wing
	5 ebony
	6 cut-wing, ebony

What was the genotype of the ebony males of the F_2?

9.　A color-blind woman married a man of normal color vision. What kind of vision could they expect their children to have?

10.　*Drosophila* females with normal (red-brown) eyes and normal (gray) bodies were mated with males having normal (red-brown) eyes and ebony bodies. The cross produced the following offspring:

Males	Females
12 wild-type	22 wild-type
10 white-eyed	19 ebony
9 ebony	
11 white-eyed, ebony	

What genotypes of the parents best account for these results?

11.　A color-blind woman married a man with normal vision. She had a daughter who, she was glad to find, had normal vision. Her daughter married a man with normal vision. What should the mother tell her daughter about the probability of having color-blind children?

12.　The mutant gene Barred in chickens is dominant and sex-linked, and the phenotype can be distinguished from normal in day-old chicks. Is there a cross that would give different phenotypes for males and females?

Chapter 9

LINKAGE AND CROSSING OVER

Alleles of genes that lie on the same chromosome tend to remain together during meiosis, a phenomenon called linkage. Gene combinations on a given homolog do not always persist through meiosis, however, because homologs of a bivalent exchange parts during Prophase I by a process called crossing over. Analysis of the recombination of many linked genes yields linear maps, on which distances between genes are measured by their probability of recombining.

A. The Behavior of Linked Genes in Random Progeny

Genetic recombination in a diploid cell undergoing meiosis requires differences in at least two genes. They may recombine in either of two ways, depending upon whether they are on the same chromosome or on different ones. We have considered the case of unlinked genes under the heading of independent assortment. Linked genes recombine by physical exchange of segments of the homologs on which they are borne. The process is called **crossing over,** and the actual site of the exchange, called a **chiasma** (Gr., crosspiece; pl., chiasmata), may be seen cytologically during meiosis. The two kinds of recombination are compared in Table 9–1.

Crosses of haploid parents *AB* and *ab* yield the progeny *AB* and *ab* (parentals), and *Ab* and *aB* (recombinants). Similarly, the diploid cross *AABB* × *aabb* produces, via gametes *AB* and *ab,* the dihybrid *AaBb.* At meiosis the dihybrid produces four types of gametes, *AB* and *ab* (parental) and *Ab* and *aB* (recombinant). The "parental" progeny or gametes have the same genotype as those that gave rise to the hybrid, and they can be identified as such in this way. Recombination is said to be **reciprocal:** there are as many *Ab* gametes as there are *aB* gametes in the example in Table 9–1. Although recombinants are fewer than parentals, each pair of alleles still segregates 1:1 in the progeny as a whole.

Table 9–1 Recombination mechanisms in eukaryotes

Mechanism	Origin	Meiotic stage	Outcome
Independent assortment	Separate alignment of two or more bivalents	Metaphase I	Equal numbers of parentals and recombinants: *AB* × *ab* yields half parentals (*AB* and *ab*) and half recombinants (*Ab* and *aB*).
Crossing-over	Chiasma between two genes on homologous chromosomes	Pachytene of meiotic prophase I	Fewer recombinants than parentals: *AB* × *ab* yields fewer *Ab* and *aB* than *AB* and *ab*. Percentage of recombinants is the **map distance.**

The criterion of linkage of two genes is that their recombinants appear in less than 50% of the meiotic products.

Crossing over occurs in Prophase I, after premeiotic DNA synthesis and the pairing of homologs. Chiasmata occur almost randomly along bivalents, but not very often. The result is that genes next to one another on a chromosome recombine rarely. Genes very far apart on a long chromosome, however, may recombine as freely as they would in independent assortment. The probability of recombination of linked genes lies between these extremes. In short distances, the frequency of recombination is roughly related to the physical distance (in DNA) between the genes involved, but, as we will see, rarely proportional to it. **The order of genes on linkage maps, however, is the same as the order of genes on the DNA.**

The geneticist uses map units, which are nothing but the percentage of recombination, to map genes. **Thus one map unit denotes 1% recombinants.** Map units are **additive** in short intervals; that is, if the genes *A, B,* and *C* were in that order with one map unit between *A* and *B,* and one map unit between *B* and *C,* recombination between *A* and *C* would be about 2%. As we will see, the additivity breaks down as distances increase, owing to multiple crossover events between genes.

B. Two-Point Crosses in Haploids

Consider the cross *AB* × *ab* of yeast, in which the *A* and *B* genes lie on the same homolog. When the diploid phase is diagrammed, a new element of notation must be introduced. The parental contributions are separated by a single line, and the two parental contributions are in the "numerator" and "denominator," respectively:

$$AB \times ab \longrightarrow AB/ab$$

This makes it clear that the haploid parents of the diploid are *AB* and *ab,* and not *Ab* and *aB.* Two pairs of terms describe the arrangement of two linked mutations in a diploid cell. **Coupling** and **repulsion** denote the arrangements *AB/ab* and *Ab/aB,* respectively. More recently, the Latin *cis* and *trans* have been more commonly used for coupling and repulsion, respectively.

If we consider a population of diploid cells undergoing meiosis, let us say in this example that the percentage of each parental gamete (*AB* and *ab*) was 45%, while that of each recombinant gamete (*Ab* and *aB*) was only 5%. If we calculate the map distance, it is 10 map units (5% *Ab* + 5% *aB*).

Let us now introduce a third gene, *C.* The *AC* × *ac* mating yields 20% recombinants. This tells us that *C,* like *B,* is linked to *A.* We know *C* cannot be between *A* and *B,* because *C* is farther from *A* than *B* is. But we cannot make a map, because we do not know which side of *A* and *B* the *C* gene lies on. Looking systematically at the three possible gene orders, we have *A-B-C, A-C-B,* and *B-A-C.* Think of the three choices as a decision of which gene is in the middle and realize that the reverse of each order is formally the same as that given. Our choices are pictured below:

(a)   *A* — 10 — *B* — 10 — *C*   (Compatible with the data.)

(b)   *A* — 20 — *C* ———— *B*   (Impossible; *C* can't be within
        |————10————| the 10-unit *A-B* interval.)

(c)   *C* — 20 — *A* — 10 — *B*   (Compatible with the data.)

The only way we can choose between (a) and (c) is to make a cross involving *B* and *C.* When we do the cross *BC* × *bc,* it yields about 10% recombinants. This eliminates the order *C-A-B*, and proves *A-B-C* correct. This example is arbitrary and could have come out the other way.

To summarize, in recombination of two genes, each with two alleles:

- Two equally frequent parental types and two equally frequent recombinant types form. The members of each class are complementary pairs. Segregation of the alleles of each gene remains 1:1.

- Map distances, in map units, are the percentages of recombinants among total progeny.

- "Two-point" (two-gene) crosses, such as the ones above, can be used to make a map. Three crosses must be done. The ends of the interval are defined by the two genes yielding the greatest number of recombinants. The other gene lies between these "end markers."

- The map is additive in short distances: the sum of map units of contiguous intervals approximates the map distance determined in the cross involving the end markers.

- Something not illustrated above is that the same map distance will be found whether the mating is *AB* × *ab* or *Ab* × *aB*. In the latter case, *Ab* and *aB* will be the parental genotypes among the progeny, but the percentage recombination between *A* and *B* will still be about 10%.

C. Two-Point Crosses in Diploids

True-breeding parents are used to begin linkage tests of diploids. In the mating of *AABB* × *aabb*, the fractional notation for the F_1, symbolized *AB/ab*, makes clear that the parental contributions are *AB* and *ab*. The F_1 will yield parental (*AB* and *ab*) and recombinant (*aB* and *Ab*) gametes in unequal numbers. **This is revealed in a testcross.** The sequence of the two crosses required is described below.

A particular cross, *AB/AB* × *ab/ab*, gave rise to an F_1 (*AB/ab*) which was then testcrossed with *ab/ab*. The genotypes, phenotypes, and percentages of this testcross progeny are:

AB/ab (*A_B_*)	43%
Ab/ab (*A_bb*)	7%
aB/ab (*aaB_*)	7%
ab/ab (*aabb*)	43%

The first and last categories (43% each) are from F_1 parental gametes *AB* and *ab;* the others (7% each) are from recombinant gametes *Ab* and *aB*. To be sure you understand this, diagram the intermediate gamete formation and fertilizations systematically.

The genes *A* and *B* are 14 map units apart. If two other series of crosses are done, starting with parents *AACC* × *aacc* and parents *BBCC* × *bbcc*, we can determine the *A-C* and the *B-C* distances. If we find *A* and *C* 6 units apart, and *B* and *C* 8 units apart, we would know that *A* and *B* are the genes farthest apart, and *C* is between them.

F_2s are relatively uninformative if they involve linked genes. Testcrosses are clearly more suited to measuring linkage in diploids. You should spend a moment diagramming the appropriate crosses to see why.

D. What Happens Between Linked Genes in Meiosis?

We may now relate the formation of recombinants, seen above, to the behavior of homologous chromosomes in a bivalent. This requires a detailed look at Prophase I of meiosis (Fig. 9–1). First, remember that **DNA synthesis (premeiotic S phase) precedes Prophase I.**

1. Stages of Prophase I

Prophase I of meiosis is quite different from prophase of mitosis. It has five stages: leptotene, zygotene, pachytene, diplotene, and diakinesis (Fig. 9–1). **Pachytene** is the stage at which the cutting and splicing of DNA in the crossing-over process takes place. The stages of Prophase I are listed below, with the major events and structures noted.

a. Leptotene.   Chromosomes appear as tangled filaments as they begin to condense.

b. Zygotene.   Homologs, already divided into two chromatids in premeiotic S phase, undergo **synapsis** or pairing, to form bivalents. Their ends remain attached to the nuclear membrane. Synaptonemal complexes (see below) begin to form.

c. Pachytene.   The homologs of each bivalent, shorter and thicker now, are held together from one end to the other by the ribbon-like synaptonemal complex, a proteinaceous structure associated with intrachromosomal recombination. Recombination of parts of non-sister, homologous chromatids takes place in the synaptonemal complex. The ends of the chromosomes remain attached to the nuclear membrane.

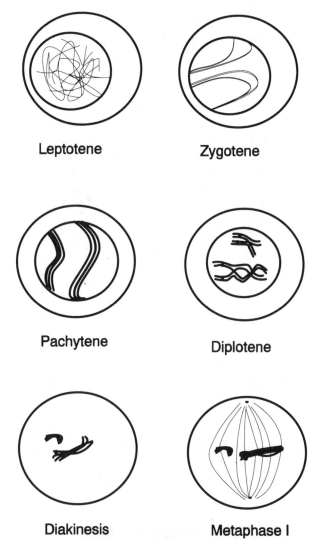

Figure 9–1   The five stages of Prophase I of meiosis, followed by Metaphase I. Chiasma formation takes place during Pachytene, when the synaptonemal complex (not shown) is fully developed (*see* Figure 9–2).

d. Diplotene. The synaptonemal complex largely disappears, remaining only at points of recombination. The bivalents continue to shorten. The "cross" structures at the points of recombination, the chiasmata, are often visible microscopically.

e. Diakinesis. At this stage, chromosomes achieve their greatest compaction. In most organisms, the few chiasmata that have formed during pachytene are important in holding the homologs together, and this persists until the end of Metaphase I. Chromosomes are also held together by other weaker mechanisms, and the lack of chiasmata in a bivalent does not normally lead to detachment of homologs. At diakinesis, the nuclear membrane breaks down.

2. Visualizing crossing over

The formation of a chiasma is illustrated at the top of Figure 9–2. Notice that sister-segments of chromatids do not change position until the homologs are separated at Anaphase I. It is useful to be able to visualize the process quickly, as in the "shorthand version" of Figure 9–2. This figure depicts unrecombined homologs in a bivalent, with a vertical line connecting the chromatids that exchange parts. You can derive the recombinant meiotic products by moving your eye along each chromatid **from the centromere** to the verticals, if any, and making the appropriate switch to the other homolog. The meiotic products are defined by whatever is connected, after crossing over, to the final four centromeres. The occurrence of recombinant meiotic products does not disturb the 2:2 segregation of alleles, and this is a check on whether your derivation of meiotic products is accurate. Get some practice deriving products from the initial bivalent, carrying mutations and marked with one or more chiasmata (as vertical lines).

3. The rules of crossing over in bivalents

Crossing over is a regular and accurate process. From genetic and cytological observations, geneticists have inferred the following behavior of bivalents in meiosis:

- Bivalents consist of two paired homologs. Each homolog consists of two chromatids. (Chromatids are often called strands, and bivalents are "four-stranded." This will cause some confusion when you realize that each chromatid is made of a double helix of DNA, each nucleotide chain of which is often called a strand.)

- Crossing over happens at Pachytene.

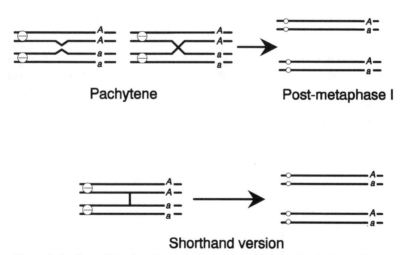

Figure 9–2 Formation of a chiasma in the pachytene stage of meiosis I, with a shorthand version of the same event. Note that chromatid segments do not change positions when a chiasma is formed; only the connections of the distal segments (farthest from the centromere) with the centromeres have changed.

- Crossing over involves non-sister chromatids.

- Any chiasma involves only two of the four strands of a bivalent.

- More than one chiasma may occur in a bivalent.

- In most organisms, chiasmata tend to inhibit the formation of other chiasmata nearby. This is called, chromosome or **chiasma interference.** The inhibition diminishes with distance and is rarely complete, even in short distances.

- If two chiasmata form in one bivalent, the strands involved in the first chiasma do not influence the choice of strands in the second. Geneticists say, therefore, that there is no **chromatid interference.**

A chiasma between a gene and the centromere of its chromosome (Fig. 9–2) will defer segregation of the alleles to the second division. The group of meiotic products will, in the fungus *Neurospora,* display this second-division segregation in the order of spores in a linear ascus, a long sac that holds the meiotic products. Instead of the order *A A a a,* the meiotic products will have the order *A a A a* or *A a a A.* Because second-division segregation results from recombination between a gene and its centromere, the frequency of the event in a population of tetrads can be used as a measure of the "centromere distance" of a gene. The underlying events are exactly the same as those leading to the recombination of two genes.

4. Multiple crossing over

More than one chiasma can occur in a bivalent. We can predict, using the short-hand method of switching chromatids at each vertical as the eye moves to the right (away from the centromere), situations in which a second chiasma cancels the recombinational effect of the first.

Double crossing over thus leads to an underestimate of map distance in longer distances, because some chiasmata are not detected as genetic recombinants. For that reason, geneticists measure long distances as the sum of shorter distances in which double crossing over is infrequent. By adding a third marker, *D/d* to the bivalent above, we can detect each of the two chiasmata as recombinants in one or the other of the two regions defined by the markers.

Figure 9–3 shows the outcome of various numbers of chiasmata between two genes. If no recombination takes place between linked genes *A/a* and *B/b*, we call the resulting tetrad a **parental ditype.** If a single crossover takes place in pachytene, only two homologs become recombinant. This yields a **tetratype** tetrad, in which only half the products are recombinant (in the cross *AB × ab,* it has *AB, Ab, aB,* and *ab* meiotic products). In order to get a **non-parental ditype** tetrad (i) two crossover events must take place, and (ii) they must involve different pairs of chromatids, so that all products are recombinant. The low probability of both (i) and (ii) occurring together makes the frequency of non-parental ditypes a good indicator of linkage of two genes. It is useful to recall that a non-parental ditype can occur as frequently as a parental ditype in the case of independently assorting genes.

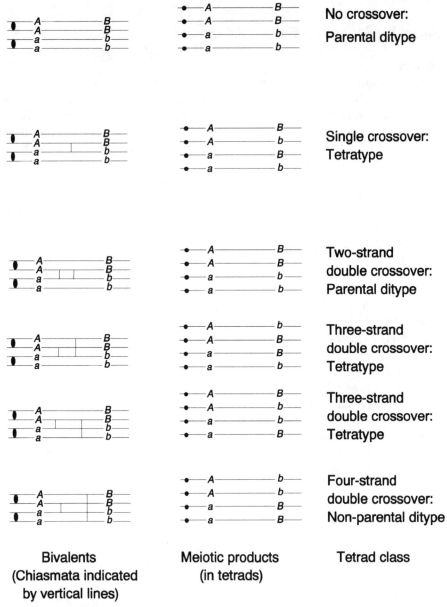

| Bivalents (Chiasmata indicated by vertical lines) | Meiotic products (in tetrads) | Tetrad class |

Figure 9–3 Tetrads emerging from bivalents with zero, one, and two chiasmata between genes *A/a* and *B/b*.

Figure 9–3 also illustrates the point that in a double-crossover bivalent, chiasmata can involve any two non-sister chromatids. There is no effect of one chiasma on the strands used in the second; as we said above, there is no chromatid interference.

Calculation of map distances in tetrads requires determining the percentage of recombinant meiotic products. It is easy to be confused between this and the percentage of tetrads of different types. Only one-half the chromatids in a tetratype tetrad are recombinant meiotic products. Therefore, barring double crossing over, the percentage of recombinant meiotic products (and thus the map distance) is only one-half the percentage of tetratype tetrads in the cross.

The measurement of linkage distances began long before the sophisticated view of tetrads described above was developed. Accordingly, as we will point out below, it is wholly unnecessary to know how recombinants arose in tetrads when we map genes. All that matters in measuring map distance is the knowledge of how many recombinant meiotic products form in a cross.

E. Mapping Genes with Three-Point Crosses

1. General

Genes are mapped by measuring the frequency of recombinant meiotic products. We simply cross haploids, or we testcross hybrid diploids, and count recombinant meiotic products. With sufficient numbers of progeny, we have more confidence in short map distances, because multiple crossing over is rare in short intervals. In a three-point cross, we measure the length of two regions at once, and determine the degree of chromosome (or chiasma) interference. The latter cannot be measured with two-point crosses. A three-point cross is illustrated in the next section with the methodology of analysis.

The cross $ABC \times abc$, regardless of the order or linkage of the genes, will give rise to eight (2^3) different genotypes. In haploids, the frequency of the meiotic products can readily be used to determine map distances. In diploids, we apply the same logic to testcrosses of a triple heterozygote ($ABC/abc \times abc/abc$). The phenotypes of the diploid testcross progeny correspond exactly to the genotypes of the gametes (meiotic products) of the heterozygote. This allows us to use a convenient shorthand. For example, if the triple heterozygote produces a gamete AbC, the testcross offspring will be $AabbCc$ in genotype and $A\_bbC\_$ in phenotype. However, we can represent the phenotype as "AbC" (dominant character for A and C, recessive for B), which is in fact the genotype of the gamete from the trihybrid that gave rise to it.

2. Arranging the data

In analyzing a three-point cross, we wish to determine four things in the following order: (i) **gene arrangement;** (ii) **gene order;** (iii) **map distances;** and (iv) **the coefficient of coincidence,** an estimate of interference. The data from a three-point testcross are given below, with the understanding that the gene order and the arrangement of recessive alleles on the two homologs of the heterozygous parent is not known at the outset. These unknowns are signified by parenthesizing the parental genotypes and omitting fractional notation.

Parents: ($AaBbCc$) $\times$ ($aabbcc$)

Number and phenotypes of progeny (using gene symbols)

ABC	1	*aBC*	30
ABc	415	*aBc*	50
Abc	29	*abC*	426
AbC	49	(Total is 1000)	

These data may confuse you at first. To use them for the analysis, you must reorder them using two principles: (i) for three genes, there are eight possible genotypes; and (ii) parental and crossover classes appear in complementary pairs, whose members are expected to be in equal numbers. **If there are less than eight pairs, the missing one(s) must be identified and listed.** The most dependable method of doing so is to write down the four genotypes that all have the *A* allele. The other four genotypes are the same, but with *a* instead of *A*. Then list any genotype that is not accounted for in the progeny, with zero individuals. After that, match the complementary pairs. Finally, arrange the data in descending order of frequency, which will correspond to parental, single crossovers (two pairs) and double crossovers, with the caution that you still do not know the order of genes:

ABc	415	*Abc*	29
abC	426	*aBC*	30
AbC	49	*ABC*	1
aBc	50	*abc*	0 (the missing class)

Occasionally, members of a complementary pair may appear in quite different numbers owing to poor viability imparted by one of the genes or gene combinations. Therefore, it is essential that pairing of recombinant classes be done first by complementary genotype, not by their similar numbers.

3. Analysis of the Data

a. Gene arrangement.   The testcross yielded *ABc* and *abC* as the most frequent classes. These must have been the "parentals," that is, progeny arising from gametes with unrecombined homologs. This reveals the arrangement of the mutant alleles: in the trihybrid, *a* and *b* lie on one homolog (in coupling, or *cis*) and *c* lies in *trans* to them, on the other: (*abC/ABc*). Only when you know this can you proceed to determine gene order.

b. Gene order.   Our goal is to identify which genes are on the ends of the array, and which one is in the middle. To do this, we must first identify the double crossover genotypes. Those are the least frequent progeny classes, or they may be absent entirely. The genotypes of the double crossover classes are then compared to the parental genotypes, which we know are the most frequent. **The end genes are identified as those that retain the same arrangement in parental and double-crossover categories.** In the example above, *AB* and *ab* are arranged the same way in both parentals (*ABc, abC*) and double crossovers (*ABC, abc*). This is not the case for the combinations of alleles of *A* and *C* or the combinations of alleles of *B* and *C*. The order is therefore *A-C-B*. We prove this by writing out the genotype of the parent heterozygote with this order and predicting what the double crossover classes would be, as follows:

We may now rewrite the genotypes with the proper gene order, and sum the members of the pairs to get the total number of recombinants in each category.

415	*AcB*	841 Parentals
426	*aCb*	
49	*ACb*	99 Single crossovers,
50	*acB*	(*A-C* region)
29	*Acb*	59 Single crossovers,
30	*aCB*	(*C-B* region)
1	*ACB*	1 Double
0	*acb*	(both regions)

Notice that we recognize which region recombination occurred in by comparing adjacent genes in parental and recombinant classes.

c. Map distances.   The *A-C* interval has 49 + 50 + 1 recombinants, for a total of 100. This is 10% of the progeny, or 10 map units. Similarly, the *C-B* interval is 6 units long ([29 + 30 + 1]/1000 × 100%). Notice that **double crossovers are used in calculating both map distances,** because we are counting crossover events, not merely progeny. The map distance between the non-contiguous *A* and *B* markers, 16 map units, is the sum of the two intervals. Now, if the *C/c* gene was not recognized in the cross, the double crossovers would not be recognized either. The result is that the map distance would be 15.8% (158/1000 × 100%). This demonstrates the effect of undetected multiple crossing over which, in the extreme, limits the amount of recombination of distant markers—no matter how far apart they are—to 50%.

d. Interference.   In testing for interference, we will determine whether one crossover event (chiasma) inhibits the occurrence of another in the same bivalent. We test for this by seeing whether crossing over in one region is independent of crossing over in the other. If one crossover completely inhibited all other crossovers on the same chromosome, no double crossovers would appear, and we would say interference was complete. If one crossover had absolutely no effect on the occurrence of another involving the same chromosome, we would say that there was no interference at all. How can we tell whether there is no interference, a little interference, or strong interference? We use the multiplication rule (Chapter 5) to answer this question. With it, we derive a number called the coefficient of coincidence, which tells us how often crossovers "coincide" with one another, yielding double crossovers.

The multiplication rule tells us that independent events occur together at a frequency that is the product of their separate probabilities. Thus, if there were no interference, the probability of double crossovers would be the product of the map distances expressed as decimal frequencies of the intervals involved. (This emphasizes that a map distance states the probability of crossing over.) If we do this with the data at hand, we would expect 0.10 × 0.06, or 0.006 of the progeny (6 out of 1000) to be double crossovers—if there were no interference.

What do we find? In the 1000 progeny, there is only one double crossover (0.001), substantially less than the 6 expected. The coefficient of coincidence (c.c.) is simply the ratio of the observed double crossovers to the expected double crossovers:

$$\text{c.c} = \frac{\text{frequency of doubles observed}}{\text{frequency of doubles expected}} = \frac{0.001}{0.006} = 0.17$$

The value for interference is 1 − c.c., or 0.83.

The steps of analyzing a three-point cross are summarized in Table 9–2.

F. Recombination at the Molecular Level

The rule of 2:2 segregation of alleles in tetrads is not absolute. Rarely, it is violated, yielding 3 *A* : 1 *a* or 3 *a* : 1 *A* segregations. This unusual phenomenon was called **gene conver-**

Table 9–2 Steps in analyzing a three-point cross

Problem	Definition	Solution
Gene arrangement	Distribution of the mutant alleles on parental homlogs	Note genotypes of the most-numerous testcross progeny (parental classes)
Gene order	Order of genes on the chromosome	End markers are those with the same arrangement in parental and double-crossover (least numerous recombinant) classes.
Map distance	Percentage of recombinants for adjacent genes	Determine percentage recombinants, using appropriate single-crossover class plus the double-crossover class.
Coefficient of coincidence	Ratio of observed double crossovers to those expected if chiasmata form independently of one another in the two regions	(i) Calculate observed double crossovers as a decimal fraction of progeny. (ii) Calculate product of map distances as **decimal frequencies.** (iii) Divide (i) by (ii). Interference is indicated by a c.c. less than 1.0.

sion, because it appeared that one of the alleles in the heterozygote had been converted, in the meiotic process, to the other. The properties of this process are:

- The frequencies of $3A : 1a$ and $1A : 3a$ segregations are roughly equal.

- Mutational sites very close to one another, most often within the same gene, are frequently co-converted.

- Conversion is associated with crossing-over of genes flanking the converted gene about 30–50% of the time, and the crossover involves the chromatids that interact in gene conversion.

Meiotic gene conversion is a rare process, and the phenomenon does not cause major distortions of progeny frequencies in nature. However, when gene conversion was discovered, many felt that it would provide insights into the formation of chiasmata. Many geneticists attempted to make a model that would incorporate both crossing over and gene conversion. The first was R. Holliday, whose model was later refined by Meselson and Radding. The steps of the recombination and conversion process that they formulated are shown in Figure 9–4. In this description, we stress that each chromatid is a double helix of DNA. The process begins with the formation of a **Holliday structure** by invasion of one chromatid by a single nucleotide chain of a non-sister (but homologous) chromatid during pachytene. This is accompanied by displacement of the complement of the invaded chromatid. The pairing of a resident and the complementary, invading nucleotide chains form a DNA **heteroduplex** (*A* in one nucleotide chain, *a* in the other) if allelic differences prevail. The Holliday structure is complete when the displaced loop is removed, and the loose ends of the nucleotide chains are ligated to one another, respecting their 5′–3′ polarities.

We now have a Holliday structure and possibly a heteroduplex segment of DNA. As we go on, you will see that **the Holliday structure is the origin of crossing over, and the heteroduplex DNA is the origin of gene conversion.** The options open to the heteroduplex and the Holliday structure are independent. The point of mismatch in the heteroduplex may be repaired (*see* Chapter 16). Second, the genes on either side of the Holliday structure, the flanking genes referred to above, may recombine. Here, more specifically, are the possibilities for the heteroduplex and the Holliday structure.

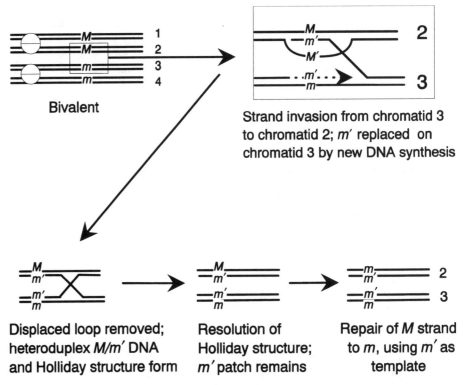

Bivalent

Strand invasion from chromatid 3 to chromatid 2; *m'* **replaced on chromatid 3 by new DNA synthesis**

Displaced loop removed; heteroduplex *M/m'* **DNA and Holliday structure form**

Resolution of Holliday structure; *m'* **patch remains**

Repair of *M* **strand to** *m,* **using** *m'* **as template**

Figure 9–4   Formation of a heteroduplex *M/m'* DNA and an associated Holliday structure, followed by gene conversion. Note that only two of the four chromatids (numbered) of the bivalent are involved.

- The heteroduplex mismatch:

    —Repair *M* to *m* (Fig. 9–4)

    —Repair *m* to *M* (In Fig. 9–4, this would restore a 2:2 ratio of alleles.)

    —No repair: **postmeiotic segregation** takes place. Replication of a heteroduplex in the first division after meiosis will yield two different, homoduplex, daughter molecules.

- Isomerization or no isomerization of the Holliday structure (Fig. 9–5)

    —No isomerization (no recombination of flanking markers)

    —Isomerization (recombination of flanking markers)

The connection between gene conversion and crossing over was appreciated when geneticists concluded that gene conversions originated in Holliday structures. The association of about 50% of conversions with crossing over of flanking markers suggested that the Holliday structure could isomerize rather freely, yielding an almost random distribution (50% of each) of recombined and non-recombined chromosomes among convertants.

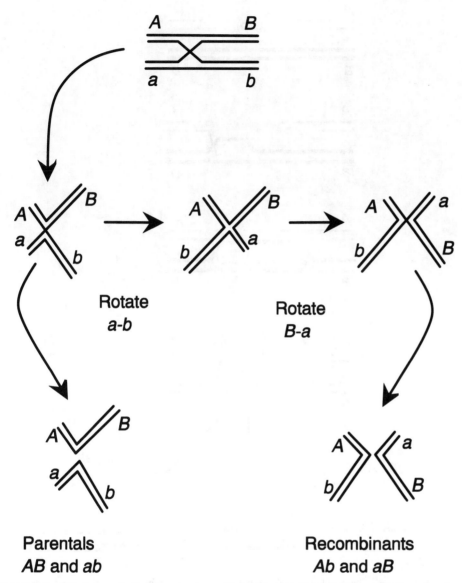

Figure 9–5 Resolution of a Holliday structure with (right) and without (left) isomerization, showing how isomerization (middle series, left to right) leads to recombination of flanking markers *A* and *B*.

More recently, another mechanism of recombination called the **double-strand gap repair model,** was proposed. It is free of some defects of the Meselson-Radding model that emerged after it was formulated. The most important differences between the two models are: (i) A two-stranded gap is repaired, using both DNA strands of the homologous chromatid as templates. This automatically leads to a 3:1 ratio (in the resulting tetrad) of alleles found within the gapped region. (ii) Two Holliday structures form, one at each end of the gap. Heteroduplex DNA may form at each one. (iii) The point of crossing-over, if it occurs, can be on either side of a conversion event. This model is diagrammed in Figure 9–6, which shows how a variety of cutting and rejoining of nucleotide chains may take place, and the consequences for the recombination of flanking genes.

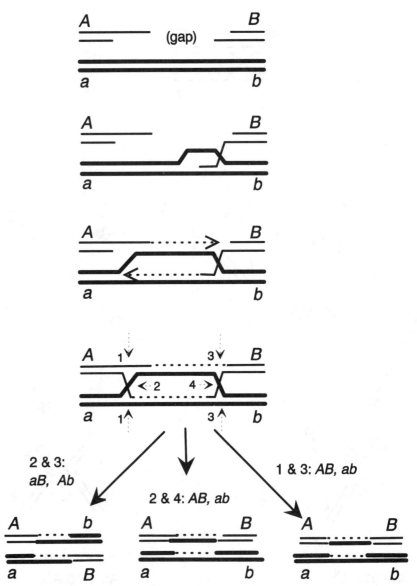

Figure 9–6   The double-strand gap repair model of gene conversion and crossing over, showing products of recombination according to different resolutions of the Holliday structures. The resolutions are shown as cutting the crossed strands as shown (arrows 2 and 4) and religating the ends after separation, or cutting at points 1 and 3 *after* isomerization. The isomerization has the effect of crossing the outer strands, and uncrossing the inner strands (*see* Figure 9–5). Only the two chromatids that interact are shown.

REVIEW QUESTIONS

1. At what stage of meiosis do the following processes take place: (i) segregation of alleles; (ii) chiasma formation; (iii) independent assortment; (iv) chromosome pairing; (v) second-division segregation?

2. What is the definition of a map unit in the study of linkage?

3. Second-division segregation in *Neurospora* was used as strong evidence that meiotic crossing-over occurred after chromosome replication. What do you think was the logic of this argument?

4.   Non-parental ditype tetrads appear in crosses involving (i) linked genes or (ii) unlinked genes. What is a non-parental ditype, and how does it form in each case? Diagram your answer (words alone may not suffice).

5.   Studies of gene conversion revealed that about half of the conversion events were associated with the recombination of flanking markers (i.e., markers on each side of the conversion event). How was this explained?

6.   How is map distance determined in (i) haploids; (ii) diploids?

PROBLEMS

1.   Three crosses involving three linked genes were performed, with the resulting "shorthand" designation of phenotypes indicated:

$AB/ab \times ab/ab$:   40% AB, 40% ab, 10% Ab, 10% aB

$AD/ad \times ad/ad$:   46% AD, 46% ad, 4% Ad, 4% aD

$BD/bd \times bd/bd$:   44% BD, 44% bd, 6% bD, 6% Bd

What is the best map of the three genes?

2.   In *Drosophila*, the eye-color mutation scarlet (*st*) and the bristle mutation spineless (*ss*) are located on an autosome 14 map units apart. What phenotypes, in what proportions, would you expect in the progeny of the following mating?

$$\frac{st^+ ss^+}{st\ ss} \text{ females } \times \frac{st\ ss}{st\ ss} \text{ males}$$

3.   In the bivalent shown here, assume that there is one chiasma between the middle two chromatids in the A-B interval and that there is another between the second and fourth chromatids in the B-D interval. What are the genotypes of the meiotic products?

4.   In a testcross of a female *Drosophila* heterozygous for three linked, recessive genes, the following phenotypes of progeny were obtained:

ABd	413	abD	423
AbD	12	aBd	12
Abd	19	aBD	20
ABD	0	abd	1

Give a map of the genes with a coefficient of coincidence.

5. In a chromosome region having the map *A*—10—*B*—20—*D*, and a coefficient of coincidence of 0.5, how many double crossovers will there be among 1000 progeny of the testcross *ABD*/*abd* × *abd*/*abd*?

6. Given the map *A*—10—*B*—20—*D*, and with no chromosome interference, how many phenotypically *ABd* individuals would you expect in a progeny of 1000 of a cross of *AbD*/*aBd* × *abd*/*abd*?

7. In maize, the genes *v* (virescent seedlings), *pr* (red aleurone), and *bm* (brown midrib) are all on chromosome 5, but not necessarily in that order. The cross

$$\frac{v^+\ pr^+\ bm^+}{v\ \ pr\ \ bm} \times \frac{v\ pr\ bm}{v\ pr\ bm}$$

produces progeny with the following phenotypes:

v^+ *pr bm*	209
v pr^+ bm^+	213
v^+ *pr* bm^+	175
v pr^+ *bm*	181
v^+ pr^+ *bm*	69
v pr bm^+	76
v^+ pr^+ bm^+	36
v pr bm	41
	Total 1000

Determine the gene order, map distance, and coefficients of coincidence.

8. In *Drosophila*, the sex-linked genes cut wing (*ct*), lozenge eyes (*lz*), and forked bristles (*f*) are in the order given and the following map distances apart: *ct* to *lz*, 7.7 units, and *lz* to *f*, 29.0 units. Assuming that there is no interference, what number of each genotype is expected if 1000 flies are recovered from the following cross? (Note the use of the "+" shorthand for wild-type alleles.)

$$\frac{ct\ lz\ f}{+\ +\ +} \times \frac{ct\ lz\ f}{\leftharpoondown}$$

9. A testcross in maize yields the following results for four loci.

abde	1	*abDE*	25
aBde	1	*abdE*	24
AbDE	1	*aBDE*	26
ABDE	2	*aBdE*	26
Abde	21	*AbdE*	444
AbDe	22	*abDe*	447
ABde	23	ABdE	455
ABDe	23	*aBDe*	459

Determine the genetic map for these four loci.

10. In maize, two linked genes, R and D, lie on chromosome 9. RRdd and rrDD plants were crossed to produce a dihybrid. The dihybrid was then self-crossed, and it produced 0.01 double-recessive (rrdd) plants. From this datum alone, can you infer how far apart the R and D genes are?

11. In a testcross of a corn plant heterozygous for three linked recessive genes (a, b, and d), the following phenotypes and numbers of progeny were obtained (total is 1000):

ABd	460	abD	476
AbD	12	aBd	12
Abd	19	aBD	20
		abd	1

Develop a map of the genes, and give the coefficient of coincidence.

12. A testcross of a trihybrid (AaBbCc) yielded the following progeny:

40 ABC		44 abc	
42 ABc		43 abC	
39 AbC		40 aBc	
37 Abc		45 aBC	

What are the linkage relations of these genes?

13. Three genes of corn, R, D, and Y, lie on chromosome 9. The map of the three genes, using map units, is

$$R \text{------} 10 \text{------} D \text{------} 20 \text{------} Y$$

In a testcross of a trihybrid plant (RrDdYy) arising from true-breeding RRDDYY and rrddyy parents, how many R_ddY_ plants would be expected in a progeny of 1000 if there were no interference? If the coefficient of coincidence were 0.3, how would this change your answer?

14. In maize, two linked genes, R and D, lie on chromosome 9. A two-point cross was done to establish the map distance between them. A dihybrid was made by mating RRdd × rrDD. The dihybrid (RrDd), crossed with rrdd, yielded phenotypes in the following frequencies: 0.4 rrD_, 0.4 R_dd, 0.1 rrdd, and 0.1 R_D_. If this particular dihybrid were self-crossed, what frequency of rrdd progeny would you expect?

15. A testcross of a trihybrid (AaBbCc) yielded the following progeny:

40 ABC		41 abc	
39 ABc		40 abC	
10 AbC		9 aBc	
8 Abc		11 aBC	

What are the linkage relations of these genes?

16. A true-breeding, white-eyed, cut-winged female *Drosophila* was mated with a true-breeding normal male. Their F_1 progeny were normal females and white-eyed, cut-winged males. Intercrossing these F_1 flies yielded the following phenotypes among the progeny:

Females	Males
90 white-eyed, cut-wing	89 white-eyed, cut-wing
98 normal	99 normal
5 white-eyed	6 white-eyed
7 cut-wing	6 cut-wing

Give the genotypes of the F_1 males and females, and determine map distances between the genes involved.

17. In a chromosome region having the map *A—8—B—20—D,* and a coefficient of coincidence of 0.4, how many progeny of the phenotype *A_bbdd* will there be among 1000 progeny of the testcross *ABD/abd* × *abd/abd?*

18. The *ad-3* gene of *Neurospora* undergoes second-division segregation in 1% of the tetrads. How far is the *ad-3* gene from its centromere?

19. A diploid of yeast heterozygous for the genes *A/a, B/b,* and *D/d* yielded the following progeny when induced to sporulate:

ABd	467
Abd	30
AbD	1
aBd	4
aBD	23
abD	475

Make a map of the three genes, giving gene order, map distances, and coefficient of coincidence.

20. A new yeast mutant, *arn1,* was found to be linked to the genes *tup4* and *spd1* on chromosome XV. In the course of this, it was found that haploid strains carrying both *arn1* and *tup4* were unable to survive. It was necessary to map the genes, and a *spd1 tup4* double mutant was mated to an *arn1* mutant. The numbers of progeny were as follows:

wild type	6
arn1	78
tup4	13
arn1 spd1	10
spd1 tup4	81

Give a map of the three genes, with map distances, as best you can calculate them.

21. A double-mutant strain containing unlinked mutations is mated to wild-type. Each gene lies very close to its centromere, such that no second-division segregation takes place. In this cross, what are the expected frequencies of parental ditype, non-parental ditype, and tetratype tetrads?

Chapter 10

CHROMOSOMAL VARIATION AND SEX DETERMINATION

A large range of genetic and phenotypic effects is associated with abnormal chromosome structure and number. Variation in chromosome number reveals the importance of gene balance and in the case of the sex chromosomes, has given great insight into sex determination. Variation in the arrangement of genes in chromosomes does not affect the phenotype unless it disrupts, removes, or duplicates some sequences. However, the genetic effects of rearrangements may be profound, owing to the difficulties in chromosome pairing and distribution at meiosis.

Chromosomal variations are easily classified, but the origins, the effects, and the genetic consequences are quite diverse. Chromosomal variations fall under two major classes (Table 10–1): changes in chromosome structure, and changes in chromosome number or **ploidy.**

A. Changes in Chromosome Structure

Changes in chromosome structure (Fig. 10–1) result from breakage and improper rejoining of chromosome fragments or from illegitimate recombination events. We often call such changes **macrolesions.** X-rays are the most common experimental method of producing chromosome breaks, owing to the energy of their ionization paths. The broken ends of chromosomes, because they lack telomeres, have an unusual tendency to rejoin with other broken ends, and this may culminate in chromosomal rearrangements rather than restoration of the original chromosome structure. Often the break-points are within genes, and therefore specific phenotypic alterations may accompany chromosomal rearrangements.

Deletions (often called **deficiencies**) are macrolesions in which genetic material has been removed from a chromosome. Deletions are often lethal, even in heterozygous form, owing to loss of vital genes or to gene imbalances. This may be true of very small deletions, particularly in haploid organisms, suggesting that genes are close together, and many of them have indispensable functions. However, in some genetically well-known species, notably *Drosophila,* use has been made of small deletions to map very small areas of chromosomes. Such deletions are often viable, if not wholly normal, in heterozygous form. Consider a heterozygote in which one homolog has a structurally normal chromosome bearing a recessive mutation, and the other homolog has a small deletion of DNA that removes the gene in question. Such a heterozygote will be hemizygous for the recessive mutation and will express it phenotypically. The phenotypic "uncovering" of the recessive mutation is called **pseudodominance.** If a number of overlapping deletions are available in a chromosome region, together with recessive mutations in the same region, it is possible to

Table 10–1   Chromosomal aberrations

Type of variation	Name	Definition and origin
Chromosome number	Euploid variation	Abnormal number of **sets** of chromosomes; may follow failure of nucleus to divide after chromosome duplication, or fertilization of an abnormal diploid gamete by a normal haploid gamete. Ex.: tetraploidy, triploidy.
	Aneuploidy	Unusual number of individual chromosome(s). Failure of disjunction of a chromosome in mitosis or meiosis. Ex.: trisomy 21 (Down's syndrome).
Chromosome structure	Deletion, deficiency	Loss of chromosome segment. May occur through two breaks and loss of intermediate segment or through loss of a chromosome tip.
	Insertion	A foreign DNA fragment interrupting the normal sequence of a chromosome. Some insertions are translocations from other parts of the genome; others are mobile genetic elements called transposons.
	Duplication	Duplication of chromosome segment in **tandem** or elsewhere in the genome. May occur through unequal recombination (tandem) or by distribution of normal and translocated chromosome segments to the same pole during meiosis of cell heterozygous for a translocation.
	Translocation	Transposition of chromosome segment to elsewhere in the genome. May be **reciprocal,** in which chromosomes **exchange** parts by improper rejoining after breaks in different chromosomes.
	Inversion	Inversion of a segment of chromosome. Arises by inverting intermediate segment between two chromosome breaks.

map the endpoints of the deleted DNA and the order of the mutations by their expression in mutant/deletion heterozygotes.

Insertions into chromosomes are often translocations of DNA from elsewhere in the genome, which are covered in this chapter. Other insertions are foreign DNA, often of **mobile genetic elements** called **transposons** that have the capacity to duplicate themselves and to move from one location in the genome to another owing to the specialized enzymatic machinery that they encode. (Transposons of bacteria are briefly described in Chapter 18.)

Duplications of chromosome regions may be separated from one another, or they may be adjacent. Large duplications of chromosomal material lead to gene imbalances that may be lethal to a zygote or even, in the case of plants, to the pollen or ovule that carries them. Duplications in which the copies lie at different positions on the same chromosome or on different chromosomes may arise through matings: normal gametes may fuse with abnormal gametes in which the same chromosomal segment has been translocated to another location in the genome.

Small **tandem duplications,** in which duplicated segments lie adjacent to one another, occur frequently in complex organisms. Tandem duplications may arise from an illegitimate recombination event in meiosis where one homolog receives two copies of a chromosome segment, and the other receives none. Homozygotes for a tandem duplication can generate higher numbers of the segment through unequal crossing over in meiosis. Consider a duplication symbolized by A A′, paired with another copy of the duplication A A′. If A on one homolog pairs with A′ on the other, followed by crossing over in the paired segment, one meiotic product will receive three copies of the segment and the other only one.

Duplications have another significance. An extra copy of a gene is free to evolve through the acquisition of mutations into a gene having a more specialized or different

Standard sequence

Chromosome 1

a b c d e f g

Chromosome 2

j k l m n o

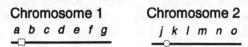

Deletion (deficiency)

a b c f g

Deletion heterozygote

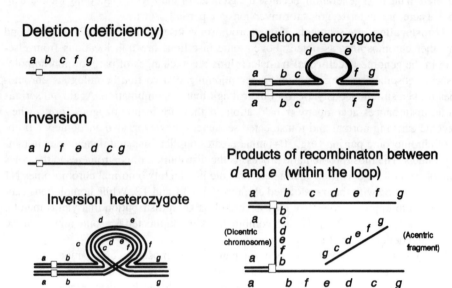

Inversion

a b f e d c g

Inversion heterozygote

Products of recombinaton between d and e (within the loop)

Reciprocal translocation:

standard:

a b c d e f g

j k l m n o

translocation:

a b c d m n o

j k l e f g

Translocation heterozygote

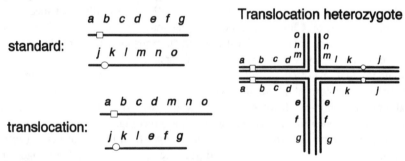

Figure 10–1 Aberrations in chromosomal structure.

function. This process does not compromise the original function, since an intact copy of the original gene remains in the genome. Duplications are therefore raw material for evolution, and many examples of duplication and specialization are known. One example is the evolution of several forms of the β chain of hemoglobin, one specialized for function in the fetus and another in the adult. In fact, it is clear that the α and β chains themselves are different derivatives of an ancestral form of an oxygen-binding protein.

A chromosome carrying a duplication or a deficiency yields a characteristic loop when it pairs at meiosis with a normal homolog. The loop represents the material that has no counterpart in the homolog (Fig. 10–1).

Inversion of a segment of a chromosome can be visualized as the difference between the normal sequence *a-b-c-d-e-f-g* and the inverted sequence, *a-b-f-e-d-c-g*. Inversions have no physiological effects if the break-points do not fall within a gene. The inverted gene order, however, results in an inverted linkage map (*see* Chapter 9) if both parents of a testcross carry the inversion. On the other hand, heterozygotes for inversions have serious problems in chromosome pairing at meiosis (Fig. 10–1), and recombination within the characteristic loop leads to chromosomes with duplications, deficiencies, and in some cases two centromeres, after recombination in meiosis. These abnormalities are usually not recovered in the next generation, because the gametes or the zygotes receiving them are inviable. Therefore, heterozygotes for inversions are partially sterile.

Translocations are segments of one chromosome inserted into or attached to the end of another chromosome; as noted above, some insertions are translocations from elsewhere in the genome. **Reciprocal translocations** are exchanges of parts of non-homologous chromosomes, arising from improper rejoining of two non-homologous chromosomes broken simultaneously, or from their illegitimate recombination. Again, no serious meiotic disturbances accompany translocations if they are in homozygous form. A heterozygote carrying normal and translocated sequences, however, also encounters a problem of chromosome pairing (Fig. 10–1) that creates duplications and deficiencies in meiotic products. The most problematic process is the distribution of centromeres to the poles in Anaphase I of meiosis. Consider a heterozygote in which two normal chromosomes N1 and N2 are paired with the translocated chromosomes T1 and T2. While homologous centromeres go to opposite poles, we may nevertheless get a distribution that yields meiotic products N1 + T2 and N2 + T1. Such combinations are duplicated for some regions of the genome and deficient for others.

Because of the selective recovery of the parental chromosome combinations N1 + N2 and T1 + T2, genes that are normally not linked appear to be linked in the progeny of a translocation heterozygote. This is handy for geneticists who wish to determine quickly the chromosomal location of a new mutation. By marking a translocated chromosome with a visible mutation, the linkage of a new mutation to the marker will show that the latter is on one of the two chromosomes of which the compound chromosome is made.

During the development of separate populations of some species, different chromosomal aberrations may arise and become common, or even pure or "fixed," in one group. Similar chromosomal variations may be distinguishing features of different species arising from a common ancestor. Chromosomal differences may be a step in the direction of reproductive isolation, since populations differing in chromosome structure may yield fewer viable progeny or progeny with poor reproductive capacity.

In some populations of *Drosophila,* persistent chromosomal **polymorphism** prevails, particularly for inversions. An inversion type may be found more commonly in one geographical area than another. However, in areas where two populations overlap, heterozygotes carrying both gene sequences arise through interbreeding. The investigators who discovered this phenomenon interpreted it as a means by which blocks of genes (within or around the inverted segments) were prevented from recombining with a homologous block of genes from the other population. The inversions appeared to be associated with a strong suppression of crossing over in heterozygotes, so that the production of inviable gametes was minimized. In this way, alleles of genes which worked well together ("coadapted") could be kept together. Some inversion heterozygotes had an even greater fitness (survival or reproductive success) than either homozygote in certain geographic areas. The latter is one form of "hybrid vigor," and the reproductive advantage of an inversion heterozygote over both homozygotes may explain the persistence of the polymorphism.

B. Changes in Chromosome Number

Changes in chromosome number may involve the increase or decrease of only one or more individual chromosomes. This is called **aneuploid variation,** as seen in trisomy (2N + 1) and monosomy (2N − 1) in a diploid organism. These abnormalities usually arise as a result of nondisjunction, a failure of individual chromosomes to separate properly in mitotic or meiotic cell division.

The other type of change in chromosome number is the increase or decrease of the number of sets of chromosomes. This is called **euploid variation,** as seen in tetraploidy (4N) and triploidy (3N). Tetraploid cells may arise by **endoreduplication,** in which chromosomes divide in prophase of mitosis, but are not separated by a mitotic spindle. The daughter chromosomes are incorporated into a single, reconstituted nucleus, with double the number of chromosomes, in an undivided cell. This problem can be induced experimentally by applying the drug **colchicine** to dividing cells. The drug inhibits formation of the mitotic spindle and induces a diploid cell, for instance, to become a tetraploid cell. Triploids may form through the fertilization of an abnormal diploid gamete, arising from a failure of a meiotic division, with an ordinary haploid gamete.

Aneuploidy and euploid variation again reveal the importance of the quantitative balance of gene activities. In plants, euploid variants such as monoploids, triploids, and tetraploids are very similar in appearance and function to the diploid from which they were derived. However, aneuploids, with gains or losses in individual chromosomes, can seriously disturb the normal phenotype, often to the point of lethality. This is true even in monosomics and trisomics of diploid organisms, neither of which actually lack any given gene entirely. The physiological effects of these imbalances are much more severe in animals than in plants. In addition, the phenotype of aneuploids is characteristic of the chromosome(s) by which they differ from the diploid. A human example is the Down's syndrome (see below), a severe developmental disorder caused by the victim having three copies of one of the smallest chromosomes, number 21. Other human trisomics are so severely affected that few embryos develop to the point of live birth. Among spontaneously aborted fetuses, more than 20% have chromosomal abnormalities, most of which are aneuploid.

With respect to genetic effects of changes in chromosome number, haploid plants, which may be phenotypically almost normal, cannot undergo the meiotic divisions at all, and triploids yield almost uniformly aneuploid gametes. This is because the three homologs of each chromosome cannot be regularly distributed to the two poles of the meiotic spindle at Anaphase I. Even some tetraploids have difficulty in chromosome pairing two-by-two, and instead form **multivalents**—Metaphase I configurations in which one homolog may pair with two others, though at different points. Aneuploids have some problems in meiosis, but only in the distribution of one or three homologs of a single chromosome.

C. Nondisjunction of Chromosome 21

Humans have 46 chromosomes (23 pairs) in their somatic cells. Forty-four chromosomes are autosomes, and the remaining two are the sex chromosomes, XX or XY. In clinical practice, the autosomes are numbered according to their length and centromere position. Thus chromosome 1 is the largest, and chromosome 22 is the smallest autosome. The X is a large chromosome, while the Y is one of the smallest of the genome. Chromosomes are visualized in the microscope after inducing cells to undergo mitosis and arresting them with colchicine at metaphase. Pictures of these "chromosome spreads" are called **karyotypes** and are used in diagnosing many genetic disorders.

Down's syndrome is an unfortunate congenital condition that confers a characteristic physical phenotype, delicate health, short life expectancy, and diminished intelligence to a child. Children with Down's syndrome have three copies, rather than the normal two, of the small chromosome 21. They usually acquire two of their three chromosomes 21 from their mothers. During the meiosis in which an abnormal egg is formed, both homologs of this chromosome go to the same pole; as a result, the egg nucleus will be disomic because it has one extra chromosome 21. The movement of both of the two homologs to the same pole, as noted above, is called **nondisjunction.** When an egg with two chromosomes 21 fuses with a normal sperm (with a single chromosome 21), the zygote will have three chromosomes 21 and will develop into a Down's syndrome child. The condition is also known as "Trisomy-21."

One mechanism proposed for nondisjunction is that two homologs of a bivalent simply fail to come apart at Anaphase I of Meiosis or when sister chromatids fail to come apart at Anaphase II (Fig. 10–2). Instead, the entire bivalent or chromosome goes to one pole. A similar event might occur in mitosis, in which two chromatids of a single chromosome are pulled to one pole of the mitotic spindle. A second result of nondisjunction is the formation of a daughter cell from the other mitotic or meiotic pole that lacks the chromosome in question entirely. If in meiosis such a cell becomes a gamete, it will yield, after fertilization, a **monosomic** zygote.

Another possible mechanism of nondisjunction, which does not exclude the first, was suggested more recently; namely, that homologs or sister chromatids disjoin prematurely. Homologs are thereby able to move independently and randomly to one or the other pole of the spindle. Both chromosomes or chromatids will therefore often go to the same pole. This mechanism of nondisjunction not only suggests a way Down's syndrome arises, but is consistent with the high correlation of the most frequent form of the disorder with the age of the affected child's mother. Girls are born with all the eggs they will produce during their reproductive lives. The eggs are arrested in late Prophase I of meiosis (often called Pro-metaphase) until they begin to be shed at the onset of puberty. When an egg is fertilized, meiosis is completed before the nuclei of the sperm and one of the meiotic products of the egg fuse. Toward the end of a woman's reproductive life, the eggs remaining will have been in Pro-metaphase I for over 40 years. Researchers conjecture that during this long interval, homologs of a bivalent occasionally fall apart before the egg is released. Thus Down's syndrome will appear more frequently in children of older mothers.

The form of Down's syndrome described above arises from spontaneous errors of chromosome distribution. Therefore, mothers who have a Down's syndrome child will not necessarily have another, although their age may predispose them to it. Similarly, the normal children of a mother who later has a Down's syndrome child do not have in-

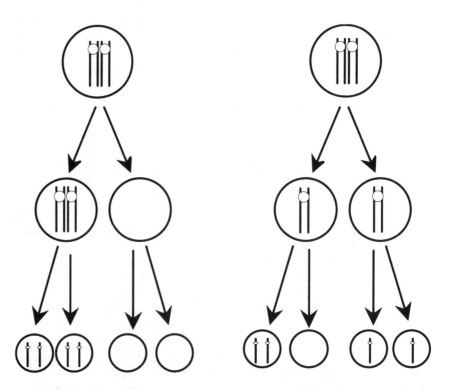

First-division nondisjunction   Second-division nondisjunction

Figure 10–2   Nondisjunction at the first (left) or second (right) meiotic division. Nondisjunction may also occur during mitosis.

creased risk of having Down's syndrome children. There is, however, a genetically persistent form of Down's syndrome arising from a translocation of much of chromosome 21 to another, larger chromosome, for instance, chromosome 14. This tends to make the attached chromosome 21 move with the large chromosome 14, regardless of how the small, normal chromosome 21 behaves. Because the translocated chromosome 21 may go to the same pole as its free homolog, the resulting gamete, when fertilized with a normal gamete, will yield a Down's syndrome child. This form of Down's syndrome is not related to the age of the mother and can arise from both abnormal sperm and abnormal eggs.

D. Sex-chromosome Abnormalities

Nondisjunction of sex chromosomes can yield abnormal disomic or **nullisomic** gametes, those with two or no sex chromosomes. A list of the simplest possibilities and the resulting zygotes is given in Table 10–2. All of these conditions have been found in humans and *Drosophila*. They give us an opportunity to ask what role the sex chromosomes have in sex determination. We shall also be able to resolve a paradox: if gene balance is so important, how do trisomics for the X chromosome, one of the largest, survive?

Sex determination in humans and *Drosophila* differs greatly. Table 10–3 illustrates this and leads us to four important conclusions:

- The Y chromosome has no impact whatever on the sexual phenotype of *Drosophila* except for sperm development. Flies with one X chromosome, but lacking a Y chromosome, are sterile but otherwise normal males. Flies with two X chromosomes are normal females, whether or not they carry a Y chromosome. (The XO grasshopper demonstrates that a Y chromosome is entirely dispensable in that species. The genes for sperm development obviously lie elsewhere in the genome.)

- In *Drosophila,* the presence of two X chromosomes confers the female phenotype. Further study shows that the critical factor is the ratio of the X chromosomes to the number of *sets* of autosomes, the so-called X:A ratio. A normal male is 1X : 2A; a normal female is 2X : 2A. The XXX female is 3X : 2A, and usually dies during pupation, a victim of gene imbalance. However, gene balance is restored sufficiently in triploid flies (3X : 3A) to yield an almost normal, viable female phenotype.

- The Y chromosome in humans determines maleness. A person lacking the Y such as the XO constitution, is phenotypically female, and a XXY person is male. These conditions are opposite in sex to the same genotypes in *Drosophila* (Table 10–3).

Table 10–2   The zygotes produced by union of normal gametes (top) and the major types of abnormal gametes arising from sex-chromosome nondisjunction (sides)

	Normal gamete		
Abnormal gamete	X egg	X sperm	Y sperm
O egg		XO	YO (lethal)
XX egg		XXX	XXY
XY sperm	XXY		
YY sperm	XYY		
XX sperm	XXX		
O sperm	XO		

Table 10–3 Sexual phenotype of *Drosophila* and humans with sex-chromosome abnormalities in *Drosophila* and humans

Organism	XO	XXY	XXX	XYY
Drosophila	Sterile male	Fertile female	Female (~lethal)	Fertile male
Humans	Sterile female (Turner's)	Sterile male (Kleinfelter's)	Fertile female (Triplo-X)	Fertile male

The XO condition is called **Turner's syndrome,** characterized by a female phenotype, but with somewhat underdeveloped secondary sex characteristics and sterile gonadal tissue. The XXY is called **Klinefelter's syndrome** with predominantly male secondary sexual characteristics, but again sterile.

- In humans, the XXX, XYY, and even XXXX, XXXXX, and XXXXXX conditions are normal in most cases, and in fact many are nominally fertile. The viability of multiple-X genotypes in humans leads us to ask how they tolerate such a large gene imbalance.

E. X-inactivation in Mammals

The **Barr body** is a condensed X chromosome found on the inner surface of the nucleus in many cells of female mammals, including humans. Study of XXY individuals also reveals a Barr body; XXX individuals have two Barr bodies, and XXXX individuals have three. In fact, all X chromosomes in excess of one become condensed and inactivated. No transcription takes place from genes on an inactive X, although the chromosome replicates and is distributed faithfully to daughter cells in cell division. "X-inactivation" in humans and many other mammals accomplishes dosage compensation, by which the "dose" of active, X-borne genes becomes equal in females and males. This hypothesis was proposed originally by Mary Lyon in the early 1960s. X-inactivation explains why humans and other mammals are so tolerant of trisomy, tetrasomy, and even higher excesses of sex chromosomes: they do not cause intolerable gene imbalances because the extra Xs are inactive. As we have seen, *Drosophila* cannot tolerate the XXX condition well, and indeed, this species accomplishes dosage compensation in another way.

Inactivation of X chromosomes in humans takes place about the 16th day in human embryonic development when there are many cells, but before the fetus is very advanced. The X chromosomes are inactivated at random. If the X chromosomes of an XX fetus differ in the alleles of a gene, half the cells will express one allele and half the other. Therefore, tissues of females heterozygous for sex-linked genes have a mosaic expression of those genes, some cells expressing one allele, other cells the other allele. Because the mosaic is so fine-grained, and because most genes are dispensable in most tissues if normal cells are nearby, genes on the X chromosome display complete or incomplete dominance relations, just like autosomal genes do.

While the mosaicism of expression of X heterozygotes is usually fine-grained, it is not always so. X-inactivation occurs before the expansion of the population of certain cell types, and thus the mosaicism is very patchy in the end. A case in point is pigment cells in the skin of mammals. Heterozygotes for alleles imparting different colors have large areas of one color and large areas of another. The tortoise-shell cat is an example of this patterning. The large-scale patchiness of coat colors was a key observation that led Mary Lyon to the X-inactivation hypothesis.

REVIEW QUESTIONS AND PROBLEMS

1. The normal chromosomes 5 and 7 of a plant have the following genetic maps (the centromeres of the two chromosomes are distinguished):

    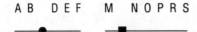

    Draw the chromosomes of (i) a reciprocal translocation of the right ends of the chromosomes, where the break points are between E and F on chromosome 5 and between O and P on chromosome 7; (ii) a deficiency for the P-R segment; and (iii) an inversion of the O-P-R segment.

2. Using the chromosome complements drawn in answer to the last question, draw the Pachytene pairing arrangements for each aberration as it would occur in a heterozygote with the normal (wild-type) sequence.

3. Draw (i) the Anaphase I configuration of chromosome 7 after a crossover between O and P in the inversion in Question 2; and (ii) the consequences of a two-strand double crossover, the first between O and P and the second between P and R.

4. How many kinds of trisomics could there be in (i) human beings; (ii) *Drosophila?*

5. Why are triploids of *Drosophila* so much more viable and healthy than XXX or other trisomic individuals?

6. Why are humans so tolerant of aberrant numbers of sex chromosomes?

7. A plant trisomic for a chromosome carrying the *A* gene has the genotype *Aaa*. What gametes would this plant produce and in what proportions?

8. What does the "Lyon" hypothesis say about the inactivation of sex chromosomes?

9. Identify the following sex chromosome constitutions as male or largely male phenotype, female or largely female phenotype, or lethal: (i) XO *Drosophila;* (ii) YO Human; (iii) XXX Human; (iv) XXY *Drosophila:* (v) XO Human.

10. The severity of Down's syndrome varies. In rare cases, two different kinds of cells can be detected in somatic tissues: normal and trisomy-21. Such cases are known as mosaics, and the ratio of normal and aberrant cells and where they occur may explain the variation of severity in this type of individual. How could mosaics like this arise?

11. What is a Barr body?

12. In *Drosophila,* XXY females are fertile. During the formation of eggs, the problem arises that three "homologs" (two Xs and a Y) must be distributed to two poles of the first division of meiosis. Geneticists found that any two of the three homologs might pair, while the third goes randomly to one pole or the other. (i) What kinds of eggs (meiotic products) would you expect an XXY female to produce according to these rules? (ii) When fertilized by a normal male, what is the sex ratio of the progeny?

13. A couple had a phenotypically normal son who, upon inspection of his chromosomes, was actually XYY in his chromosome constitution. In which parent, and at what meiotic division, could the nondisjunction causing this constitution have happened?

Chapter 11

SUMMARY OF INHERITANCE PATTERNS AND PROBLEM SOLVING

A. General

The previous chapters have explained all but one of the basic phenomena of gene transmission in eukaryotes. The exception, extranuclear inheritance, will be taken up in Chapter 14. Table 11–1 summarizes all patterns of inheritance, including extrachromosomal inheritance. The table defines each phenomenon and gives you criteria by which to recognize it in the outcome of crosses. Because there are so few fundamental phenomena, you should be able to remember them as an expanded checklist for use in problem solving. Most genetics problems will yield to formulating alternative explanations clearly enough, with proper notation, to test them with the data at hand. This is particularly true in cases of gene interaction (not included in Table 11–1, but organized in Table 7–1).

    Go through Table 11–1 carefully in order to identify any area that is unclear to you, and review its main features. This will enable you to consider the choices that problems force you to make. Problems ask you to infer the genetic basis of the number and proportion of progeny of crosses. These numbers are the data on which geneticists, from Mendel onward, have based conclusions about gametes of the parents, chromosome segregation, sex linkage, and the linkage of genes. You must have in mind not only the criteria for deciding what governs the outcome of crosses, but a systematic way—a checklist—of using these criteria. These two aids follow. **Your initial goal should be to resolve complex matters into logical components and to embody your hypotheses in consistent symbolic notation.**

B. Checklist (in priority order):

1. Is the organism haploid or diploid?

2. What alternative characters are segregating? Clarify the problem by *rigorously* formulating phenotype/genotype correlations, to the extent possible, with clear symbols. If you can't decide the relationship between phenotype and genotype (which is frequently the case), formulate *two* or more clear alternatives, specifying the difference with precision.

   a. Identify the phenotypic alternatives for each character.

   b. Assign gene symbols tentatively, if ratios are simple (*see* Table 11–1).

   c. Assign allele symbols, using modified gene symbols.

d. Determine the number of genes, if possible (*see* Table 11–1).

e. Write the genotypes of the parents, being sure each parent has one (haploid) or two (diploid) representatives of *every* gene.

f. Decide dominance relations (diploids only).

3. What is the inheritance pattern (referring to Table 11–1)?

a. Initially, consider one pair of alternative characters at a time.

b. Even if there are alternative characters, consider whether segregation occurs (extranuclear factors generally do not segregate).

c. Determine whether segregation occurs in one parent or in both parents (e.g., whether, in diploids, it is a testcross or an F_2: *see* Table 11–1).

d. Determine whether a gene is autosomal or sex-linked (diploids only).

e. Determine whether genes are independently assorting or linked.

f. Determine whether genes have independent expression or whether they interact (*see* Table 7–1).

You should return to problems at the end of preceding chapters, and go to another source of problems and see whether the checklist approach renders them easier to solve. However, complex problems are given here for you to disentangle using the checklist.

Table 11–1 Summary of gene transmission phenomena and criteria for resolving genetic transmission problems

Phenomenon	Definition	Criteria
1. Segregation	Separation of alleles of one gene in meiosis	1:1 ratio of meiotic products (haploid and diploid organisms). In diploids: 1:1 testcross ratio; 1:2:1 ratio of F_2 genotypes; 3:1 ratio of F_2 phenotypes if dominant-recessive pair
2. Independent assortment	Independent segregation of two or more genes on different chromosomes (or far apart on one chromosome)	Recombinant and parental gametes or testcross progeny are equal in frequency. Dihybrid F_2 genotypes are in a 1:2:2:4:1:2:1:2:1 ratio; F_2 phenotypes are in 9:3:3:1 ratio if both genes have dominant-recessive relationship
3. Sex linkage	Location of a gene on a sex chromosome (generally the X)	Reciprocal crosses may differ; phenotypes may distribute themselves differently by sex in a progeny
4. Linkage	Location of two genes on the same chromosome	Parental gametes (or testcross progeny) exceed recombinants with respect to any two genes
5. Extranuclear inheritance	Maternal effect, or determined by DNA of organelle or symbiont.	Non-segregating phenotype; persistence of one or the other phenotype through many generations in the case of cytoplasmic inheritance.
6. Complete dominance[a]	Heterozygote has the same phenotype as one homozygote	F_1: resembles dominant parent; F_2: 3:1 ratio of dominant to recessive phenotype
7. Semidominance or codominance[a]	Heterozygote is intermediate or shares characters of both homozygotes	F_1: intermediate phenotype; F_2: 1:2:1 ratio (corresponds to genotype)

[a] These terms refer strictly to relationships between alleles of a single gene.

PROBLEMS

1. A pure-breeding, white-eyed female *Drosophila* was mated to a pure-breeding male with cut wings and brown eyes. Their progeny consisted of normal females and white-eyed males in equal numbers. When intercrossed, these progeny produced an F_2 as follows:

    Females: 98 white-eyed, 23 brown-eyed, and 79 wild-type.

    Males: 86 white-eyed, 14 white-eyed, cut-wing, 10 wild-type, 4 brown-eyed, 21 brown-eyed, cut-wing, and 65 cut-wing.

    Rationalize this cross.

2. Two slow-growing strains of *Neurospora* were mated, and the progeny segregated as follows: 65 wild-type (normal growth) and 110 slow growing. Rationalize this cross and suggest a test of your hypotheses about any remaining ambiguity.

3. Two slow-growing strains of *Neurospora* were mated, and the progeny segregated as follows: 2 wild-type (normal growth) and 443 slow growing. Rationalize this cross and suggest a test of your hypotheses about any remaining ambiguity.

4. A man with a peculiar growth of hair on his earlobes comes from a family in which all his brothers and his father had the same characteristic. No female of his family displays the phenotype. None of the man's sisters have any children, male or female, with the trait. Rationalize this pattern of inheritance.

5. A yeast strain that had a requirement for the amino acid histidine was mated to a strain (Strain #1) that did not require histidine. The diploid grew without histidine, and when induced to undergo meiosis, the meiotic products segregated 1 His$^+$: 1 His$^-$ (here, phenotypic traits are given in Roman characters, capitalized). A mating of the histidine-requiring strain to another His$^+$ strain (Strain #2) also yielded a His$^+$ diploid, but the meiotic products of this diploid segregated 3 His$^+$: 1 His$^-$. Rationalize this cross, and suggest a test of your hypothesis.

6. Two true-breeding stocks of *Drosophila* were developed by a geneticist. One (Strain #1) was phenotypically normal in all respects, and the other (Strain #2) had normal females, but the males had very few bristles, which the geneticist called *bald*. Reciprocal matings of the two strains were made and followed through the F_2, with the results as follows:

    Strain #1 females × Strain #2 males (bald): F_1s were all normal. The F_2 males were 1/2 normal and 1/2 bald, while the females were all normal.

    Strain # 2 females × Strain #1 males: F_1 females were normal, males were bald. The F_2 males were 1/2 normal and 1/2 bald, while the females were 7/8 normal and 1/8 bald.

    Rationalize these results.

Chapter 12

POPULATION GENETICS
AND EVOLUTION

Population genetics is the application of Mendelian principles to gene frequencies in populations. A population is defined as a group of interbreeding organisms living in a geographically restricted area. Major topics in population genetics include the calculation of gene and genotypic frequencies; random mating and the Hardy-Weinberg law; and factors resulting in changes in gene frequencies.

A. Calculation of Gene and Genotypic Frequencies

The genetic information contained in a population is its gene pool, which includes all alleles of all genes in the population. For codominant alleles that produce unambiguous phenotypic effects, it is simple to calculate allele frequencies. In cattle, for example, the N and R alleles, which determine coat color, lie at a single locus. Individual cattle homozygous for NN have white coats, NR individuals have a roan phenotype (a mixture of red and white hairs), and RR individuals have red coats. In a population studied by Sewall Wright in 1917, there were 756 white, 3780 roan, and 4169 red cattle. Because the heterozygous class is distinct from either homozygous class, allele frequencies can be calculated directly:

756 white cattle:	1512 N alleles
3780 roan cattle:	3780 N alleles + 3780 R alleles
4169 red cattle:	8338 R alleles
Totals:	5292 N alleles + 12,118 R alleles

The number of alleles in the population is 5292 + 12,118 or 17,410. The frequency of the N allele is 5592/17,410 = 0.304. The frequency of the R allele is 12,118/17,410 = 0.696. Note that the sum of the frequencies of the R and N alleles is 1.0.

A simpler approach to the calculation of allele frequencies uses relative frequencies. The relative frequencies of white, roan, and red cattle are:

White	756/8705	= 0.0869
Roan	3780/8705	= 0.4342
Red	4169/8705	= 0.4789
Total		1.0000

The frequencies of alleles at a locus always sum to 1, and the frequencies of genotypes for a locus also sum to 1. Allele frequencies are calculated as follows:

$$\text{Frequency of } N = 0.0869 + (0.5)(0.4342) = 0.304$$

$$\text{Frequency of } R = (0.5)(0.4342) + 0.4789 = 0.696$$

In this approach, the contribution of the heterozygote class in each case is one half the frequency of the genotype because heterozygotes contain only one copy of each allele.

Codominant alleles affecting morphological traits are rare; to measure gene frequencies in natural populations, geneticists have turned to biochemical and molecular techniques.

Electrophoresis is used to separate proteins in a charged field in a gel or wetted paper. Once separated, the proteins are stained so that they can be visualized as bands in the gel. The patterns that emerge usually have a simple genetic basis. A band representing a protein encoded by a particular allele at a locus is known as an **allozyme.** A heterozygous organism produces two allozymes, which are codominant (*see* Chapter 6), and for this reason, electrophoretic phenotypes can be used to infer allele frequencies in populations.

Another electrophoretic method of studying genetic variation in populations is the use of restriction enzymes (*see* Chapters 1 and 20). Because the recognition sites for a given restriction enzyme appear and disappear by mutation, the length of DNA between two sites recognized by the same enzyme will vary. Organisms heterozygous for DNA fragments of different length will produce two bands on a gel when the gel is exposed to a radioactive, homologous DNA preparation capable of hydrogen bonding or "hybridizing" with the fragments (*see* Chapter 20). Homozygotes, on the other hand, will display only one band. **Restriction fragment length polymorphisms (RFLPs)** are used in the same manner as allozymes for population studies. The advantage of using DNA is the greater amount of variation visible to the investigator.

B. Random Mating and the Hardy-Weinberg Law

An understanding of the **Hardy-Weinberg law** is essential for understanding how gene frequencies in populations are governed. The law states that if mating in populations occurs at random and there is no selection, migration of genes from neighboring populations, or mutation, gene frequencies will remain constant from one generation to the next. In other words, unless something perturbs allele frequencies, we would not expect evolutionary change.

The **Hardy-Weinberg law** is understood most easily with algebraic notation. If a locus is represented by two alleles, A and a, they will have the frequencies p and q, respectively. The algebraic shorthand for writing "the frequency of A is equal to p" would be "$f(A) = p$;" consistent use of this notation will help in problem solving.

If A and a are distributed to haploid gametes, which can carry only a single copy of the gene, the frequencies of the alleles, p and q, respectively, will be identical to their frequencies among gametes. These allele frequencies define the **gene pool** of the population. **The gene pool specifies the frequencies of gametes produced by individuals, but does not specify any particular distribution of alleles in genotypes.**

Random mating occurs if gametes in the gene pool combine randomly to produce a new generation. An algebraic formulation of the Hardy-Weinberg principle is used to calculate the genotypic frequencies in the following generation:

$$(p + q)^2 = p^2 + 2pq + q^2$$

in which the values p^2, $2pq$, and q^2 are the frequencies of the AA, Aa, and aa genotypes, respectively. This formula is the expansion of the binomial expression, $(p + q)^2$, using our formal notation for allele frequencies:

$$f(AA) = p^2$$

$$f(Aa) = 2pq$$

$$f(aa) = q^2$$

The binomial expansion is based on the multiplication rule (*see* Chapter 5), which states that the probability of two independent events occurring in combination is the product of their separate probabilities. The application of this law to the Hardy-Weinberg equilibrium can also be appreciated with a Punnett square, in which values in the table are derived from frequencies of alleles in the gene pool.

Male gametes:

		p [f(A)]	q [f(a)]
Female gametes:	p [f(A)]	p^2 (AA)	pq (Aa)
	q [f(a)]	pq (Aa)	q^2 (aa)

The sum of the values in the Punnett square is $p^2 + 2pq + q^2$, corresponding to the frequencies of the genotypes *AA, Aa,* and *aa,* respectively. The Punnett square forces one to combine the gametes of the population at random in the same manner that the binomial expansion relies on the multiplication rule.

Because the effect of either foregoing method—the binomial or the Punnett square—is to distribute alleles to genotypes, the frequencies of the alleles do not change upon random mating, nor will they do so in subsequent generations. Moreover, the principle that distributes alleles to genotypes remains the same; therefore, the genotype distribution will also persist in subsequent generations. The constancy of gene and genotype frequencies from one generation to the next is referred to as the Hardy-Weinberg equilibrium.

A numeric example demonstrates the significance of a Hardy-Weinberg equilibrium. Imagine a highly artificial population in which 70% of the population (both male and female) had the genotype *AA* and 30% of the population (male and female) had the genotype *aa.* Matings in this population would be *AA* × *aa; aa* × *AA; AA* × *AA;* and *aa* × *aa. AA* individuals produce gametes carrying only the *A* allele, and *aa* individuals produce gametes carrying only the *a* allele. As a result, it is easy to see that f(A) = p = 0.7 and f(a) = q = 0.3.

We first use a Punnett square to calculate genotype frequencies:

Male gametes:

		$p = 0.7$ [f(A)]	$q = 0.3$ [f(a)]
Female gametes:	$p = 0.7$ [f(A)]	$p^2 = 0.49$ AA	$pq = 0.21$ Aa
	$q = 0.3$ [f(a)]	$pq = 0.21$ aA	$q^2 = 0.09$ aa

Summing identical genotypes, we obtain:

$$f(AA) = p^2 = 0.49, f(Aa) = 2pq = 0.42, \text{ and } f(aa) = q^2 = 0.09.$$

We may derive these frequencies more simply by using the Hardy-Weinberg equilibrium, $(p + q)^2$ or $p^2 + 2pq + q^2$, the terms of which appear in the Punnett square above, and which, like the Punnett square, assumes random mating. It will be useful for you to take this population of offspring, calculate the frequency of the two alleles (which should be the same as in the parents), and demonstrate to yourself that another generation of random mating will yield the same allele and genotype frequencies as the offspring of the original parents. **In other words, a Hardy-Weinberg equilibrium is established in a single generation of random mating.**

You should appreciate that for a dominant-recessive allele pair, the phenotype frequencies will be $f(A\_) = p^2 + 2\,pq$ and $f(aa) = q^2$. Thus only the frequency of the recessive homozygote can be determined unambiguously from the phenotypes in the population. The ability to determine the frequency of the recessive homozygote *aa* should not tempt you to assume that you know the frequency of that allele, *a,* in the population, because you do not know the frequency of heterozygotes, *Aa.* The problems at the end of this chapter develop this point more clearly.

The Hardy-Weinberg law can be extended to encompass more than two alleles of a locus. In a population in which three alleles occur at a locus, we have the following frequencies:

$$f(a_1) = p$$
$$f(a_2) = q$$
$$f(a_3) = r$$

At equilibrium, the frequencies of the genotypes drawn from this gene pool would be, after one generation of random mating:

$$(p + q + r)^2 = p^2\,[a_1a_1] + q^2\,[a_2a_2] + r^2\,[a_3a_3] + 2pq\,[a_1a_2] + 2pr\,[a_1a_3] + 2qr\,[a_2a_3]$$

In subsequent generations of random mating, the frequency of a_1, a_2, and a_3 remain unchanged.

The frequencies of alleles carried on the sex chromosomes are also governed by the Hardy-Weinberg principle. In organisms in which females are XX and males are XY, a sex-linked trait will appear in different frequencies in the two sexes. For a sex-linked recessive allele, because it is expressed as a single copy in males, it will appear in a frequency equal to the allele **frequency, q in males.** Females, on the other hand, have two copies of the X chromosome, which means that the Hardy-Weinberg formula must be used to calculate expected genotypic and phenotypic frequencies. A recessive trait will appear among females according to the *square* of the allele frequency. Thus for a locus on the X chromosome where A is dominant to $a,$ and $f(A) = p = 0.9$ and $f(a) = q = 0.1$, 10% of males would display the trait, but only 1% ($q^2 = 0.01$) of the females would do so. Two well-known human traits, color blindness and hemophilia, show this pattern of inheritance. When alleles are sex-linked, more than a single generation of random mating is required before a Hardy-Weinberg equilibrium is attained.

C. Factors Causing Changes in Gene Frequency

Evolution may be defined in genetic terms as a process of change in allele frequencies. If alleles at all loci were maintained in Hardy-Weinberg equilibrium, evolution would not occur.

The major forces that disturb this equilibrium and lead to evolutionary change are **mutation, selection, migration, nonrandom mating, and random changes in gene frequencies.** Acting singly or together, these forces have driven the vast diversification of living organisms.

1. Non-random mating

Inbreeding, a common form of nonrandom mating, refers to matings between relatives. Inbreeding leads to a decrease in the frequency of heterozygotes. The most-intense form of inbreeding is **self-fertilization,** which may occur in hermaphroditic organisms such as Mendel's peas. With each generation of self-fertilization, the frequency of heterozygotes is halved. After a few generations of self-fertilization, few heterozygotes persist in a population.

The **inbreeding coefficient,** F, is used to measure the level of inbreeding in a population. To calculate F, the observed frequency of heterozygotes, H_1, is compared to that expected, $2pq$, on the basis of the Hardy-Weinberg law using the formula:

$$F = (2pq - H_1)/2pq$$

As the observed frequency of heterozygotes decreases, the value for F approaches 1, its maximum value. At this point, there are no heterozygotes in the population.

Inbreeding is usually harmful, because recessive, deleterious alleles become homozygous. Some plant species are exceptional in undergoing self-fertilization regularly, and additional inbreeding has little effect. For species with a history of regular inbreeding, deleterious alleles have often been eliminated from populations or reduced in frequency by natural selection. For most species, however, inbreeding produces deleterious progeny genotypes, and therefore mechanisms promoting outcrossing are favored by evolution. One mechanism promoting outcrossing in flowering plants and fungi is self-incompatibility, which prevents both self-fertilization and matings between individuals of the same incompatibility group. Incompatibility reactions are governed by alleles at the S locus; the alleles of this locus may be quite numerous.

2. Mutation and recombination

Mutation provides the basic variability in populations permitting evolutionary change, and recombination greatly increases variation by generating many combinations of alleles of different genes, all of which potentially interact in their effect on fitness.

Small point mutations, such as base-pair substitutions, can have major effects on organisms. For instance, a particular base-pair substitution in the gene coding for human hemoglobin results in defective hemoglobin, called **sickle-cell hemoglobin.** This form of the molecule has reduced oxygen-carrying capacity. This leads to severe anemia in people homozygous for the defective allele, Hb^S.

With the exception of polyploidy (*see* Chapter 10), larger chromosomal changes, such as inversions, deletions, and duplications are increasingly deleterious as gene balance is increasingly disturbed. The observation that large mutational changes are generally deleterious explains why mutation cannot by itself result in adaptive evolutionary change. Rarely, however, beneficial mutations favored by natural selection or other evolutionary forces can spread throughout a population.

To appreciate the importance of recombination, consider a diploid organism having only three pairs of chromosomes. For simplicity, let us look at segregation patterns for heterozygotes of loci A, B, and C, each gene lying on a different chromosome. As we saw in Chapter 5, a particular genotype, say AbC, is 1 of 8 possible among the gametes of the triple heterozygote $AaBbCc$. The general formula for obtaining the number of gametic combinations is 2^n, in which n is the number of chromosome pairs. For the case of 3 chromosome pairs the number of gametic combinations is $2^3 = 8$. For humans having 23 pairs of chromosomes, the number of gametic (haploid) combinations arising by independent assortment is 2^{23}, or 8,388,608. Human beings have two copies of each chromosome, which means that the number of zygotic (diploid) combinations will be:

$$2^{23} \times 2^{23} = 2^{46} = 7.04 \times 10^{13}$$

These calculations do not include recombination arising through crossing over between homologs. When this process is included, the number of gametic and zygotic combinations becomes unimaginably large.

Mutation and recombination result in the abundant genetic diversity essential for evolutionary change. We can now understand the importance of sexual recombination to evolution. In organisms that cannot form recombinants because they reproduce asexually, genetic diversity is considerably reduced relative to sexual species. Some of these asexual species, in fact, may be evolutionary "dead ends" that have impaired ability to cope with rapidly changing environments. They may suffer extinction before their sexually reproducing counterparts.

3. Migration

Migration of new alleles into a population is an additional means for increasing genetic variability in the population. The significance of migration depends on its magnitude and the frequencies of genes in the source and target populations. If gene frequencies are similar in the two populations, even substantial migration of organisms between the populations will have little effect upon gene frequencies.

The allele causing sickle-cell anemia in humans provides an example of migration. This allele occurs in high frequency ($Hb^S = 0.24$) in areas such as Africa where malaria is common because Hb^S, when heterozygous, confers resistance to malaria. The frequency of Hb^S is far lower (0.048) in African-Americans in the United States because of the migration of normal alleles (Hb^A) from the white population to the black population (depending on how we categorize race, we could just as well say that the frequency of Hb^S has increased in the white population owing to migration of Hb^S alleles from African-Americans). The absence of malaria in the United States also accounts for much of the reduction in the frequency of Hb^S, because the Hb^S allele confers no advantage in malaria-free environments and imparts a serious anemia when it is homozygous.

4. Selection

a. Selective coefficients.   Early Darwinians viewed natural selection as a process of differential mortality. Modern definitions are more inclusive and recognize that selection can occur through any force that governs the genetic contribution of one generation to the next. Such forces may be found at various levels, ranging from populations of organisms to alleles of individual genes. In our discussion, we will be most concerned with selection operating at the level of individual organisms or single genes. **A modern definition of natural selection is the differential, nonrandom reproduction of alleles.**

Selection is embodied quantitatively in the selection coefficient, s. This factor represents the disadvantage of a genotype or an allele during reproduction. If $s = 1$, selection has its greatest intensity, and a given genotype would itself be inviable, or would leave no viable or fertile offspring. The Darwinian fitness of such an individual, symbolized as w, would be 0. Conversely, when $s = 0$, w would be 1. The selection coefficient, s, and Darwinian fitness, w, are related simply: $w = 1 - s$.

The advantage of these terms is that selection coefficients (or Darwinian fitness) can often be quantified by comparing the performance of two genotypes in a single environment or by comparing the fitness of a single genotype in two different environments. These comparisons show that fitness is a relative concept. Fitness can be defined only by comparing genotypes with one another or by looking at the effect of environmental differences on the same genotype. For this reason, it is meaningless to attempt a definition of fitness unless the environment is specified.

b. Selection against a dominant allele. The effect of selection at a locus initially in Hardy-Weinberg equilibrium can be represented symbolically. Let us say that in a population, alleles at the A locus are initially in Hardy-Weinberg equilibrium, and the A allele is dominant. As we have seen, the genotype frequencies at equilibrium would be p^2 [f(AA)], $2pq$ [f(Aa)], and q^2 [f(aa)]. Selection against the dominant genotype can be represented by modifying the frequencies of genotypes carrying the A allele. We use the same value for s (s_1) because of the dominance of A:

$$p^2(1 - s_1) \, AA + 2pq(1 - s_1) \, Aa + q^2 \, aa$$

Simple algebra, not presented here, shows that, when selection is against a dominant allele, it will be lost from the population if mutation does not reintroduce it. The magnitude of the selection coefficient will determine the rate of loss of the allele.

c. Selection against a recessive allele. Selection against a recessive allele can be represented similarly:

$$p^2 \, AA + 2pq \, Aa + q^2(1 - s_1) \, aa$$

(Use of the same subscript for the selection coefficient does not imply that the intensity of selection in the two examples is the same.) Simple algebra can show again that selection against a recessive allele will result in its eventual loss from a population. In comparison with selection against a dominant allele, however, the rate of loss of a recessive allele is very slow, and, for practical purposes, it may be impossible to eliminate the allele, no matter how intense the selection. Two examples of populations at Hardy-Weinberg equilibrium make this clear:

**Genotypic frequencies
at equilibrium**

	Gene frequencies	p^2 (**AA**)	$2pq$ (**Aa**)	q^2 (**aa**)
Case 1:	f(A) = 0.9	0.81	0.18	0.01
	f(a) = 0.1			
Case 2:	f(A) = 0.99	0.9801	0.0198	0.0001
	f(a) = 0.01			

For Case 1, the frequency of a alleles in aa homozygotes is 0.01, and the frequency of a alleles in heterozygotes is 0.09 (0.18/2), nine times more. However, for Case 2, similar calculation shows that the a allele is 99 times more frequent in heterozygotes as in homozygotes. Selection cannot eliminate a alleles in heterozygotes because they are recessive. As a recessive allele becomes rarer in a population, a greater proportion of these alleles are sheltered in heterozygotes, and selection will operate with decreased efficiency. As this happens, the **rate of loss** of the recessive allele will become so low that it appears to be stable in frequency. In fact, at very low levels, the Hardy-Weinberg formula may approximate the genotypic distribution adequately.

The most important conclusion of this exercise is that artificial selection is extremely inefficient in eliminating even lethal, recessive alleles from a population. This reasoning discredits attempts to improve the human race through selection against individuals homozygous for recessive, deleterious alleles.

d. Selection against both homozygous genotypes. When selection occurs against both homozygous classes (implying that dominance is lacking or incomplete), we have the following modification, after selection has had its effect, of the Hardy-Weinberg equilibrium: $p^2(1 - s_1) A_1A_1$, $2pq A_1A_2$, and $q^2(1 - s_2) A_2A_2$. The use of different subscripts for the selection coefficients operating against the two homozygous classes implies that the values may be different. Selection against the homozygotes leads to a stable equilibrium in which both the A_1 and the A_2 alleles are maintained in the population. The allele frequencies are determined by the selection coefficients:

$$f(A_1) = p = s_2 / (s_1 + s_2)$$
$$f(A_2) = q = s_1 / (s_1 + s_2)$$

Selection against both homozygotes at a locus results in heterozygote superiority. Heterozygote superiority produces an equilibrium of allele frequencies, but the genotypes in the population are not in Hardy-Weinberg equilibrium. The Hb^S allele, encoding sickle-cell hemoglobin, is maintained by selection favoring heterozygotes living in areas with malaria. In Africa, heterozygous individuals are resistant to malaria and consequently have higher fitness than either homozygote class. Individuals homozygous for Hb^A, the normal allele, may succumb to malaria, and individuals homozygous for Hb^S have a debilitating anemia.

5. Random processes in populations

When populations consist of a small number of individuals, and selection coefficients are very small or nonexistent, random events may lead to changes in frequency. The term genetic drift designates changes in gene frequency due to random effects. [Genetic drift and gene migration (or gene flow) are distinct phenomena and should not be confused with one another.] The role of population size in determining the importance of random events in populations is illustrated in the following example, in which identical events occur by chance in two populations of different sizes. In both populations, the initial gene frequencies are the same [$f(A_1)$ = $f(A_2)$ = 0.5], and the populations are in Hardy-Weinberg equilibrium at the outset:

	Genotype frequencies		
	f(A_1A_1)	**f(A_1A_2)**	**f(A_2A_2)**
Case 1 ($N = 10000$)	2500	5000	2500
Case 2 ($N = 100$)	25	50	25

The random event that occurs in both populations results in loss of 25 A_1A_1 individuals and 25 A_1A_2 individuals. The effects of this accident on gene frequencies in the following generation are to reduce the frequency of A_1 in Case 1 from 0.5 to 0.4987, and in Case 2 from 0.5 to 0.25. Thus identical accidents have radically different effects, depending upon population size. The principle illustrated here, that an accident has its greatest effect in a small population, applies universally in population genetics. Examples of changes in allele frequency known to result from drift are rare because of the difficulty in eliminating selection as an alternate explanation. Modifications of frequencies of the ABO blood alleles following the migration of small religious sects from Europe to the United States, however, provide some of the best corroborated examples.

D. Solving Population Genetics Problems

1. Categorizing the problem

All population genetics problems are concerned with the presence or absence of allelic and genotypic equilibria. Allele and genotype equilibria are the essential features of a Hardy-

Weinberg equilibrium. Cases in which only allelic frequencies, but not genotype frequencies, are in equilibrium are also possible as a result of certain kinds of selection. Population genetics problems are either of two types. In the first kind of problem, information is provided on genotypic frequencies. You are expected to calculate allele frequencies and determine whether a Hardy-Weinberg equilibrium exists. In the second type of problem, you must assume the presence of a Hardy-Weinberg equilibrium in order to determine gene frequencies. Random mating and forces resulting in changes in allele frequency are assumed to be negligible.

2. The checklist

- What are the alleles and their symbols? Use standard notation for frequencies of alleles and genotypes. Note that it may not be possible to specify the frequencies of all genotypic categories, because of dominance relationships. The categories you specify should sum to one.

- Have you enough information to test for a Hardy-Weinberg equilibrium? Only if all genotype frequencies are known can you calculate allele frequencies and test for a Hardy-Weinberg equilibrium. A match between expected and observed genotype frequencies in a single generation is necessary, but not sufficient as a criterion of the Hardy Weinberg equilibrium. As one of the problems shows, selection may result in allele and genotype frequencies that deceptively conform to a Hardy-Weinberg equilibrium. In such cases, determinations of genotype frequencies at several points in the life cycle are needed to reveal the operation of selection.

- Must you assume the presence of an equilibrium to solve the problem? If only the frequency of the recessive genotype is specified, you must assume that the population is in Hardy-Weinberg equilibrium in order to proceed with calculations. The conditions for Hardy-Weinberg equilibrium are violated when selection occurs against the recessive genotype, unless it is very rare.

- What is the criterion of conformance between observed and expected genotype frequencies, based on the Hardy-Weinberg law? Here, the use of the chi-squared test is appropriate, but recognize the need to use actual numbers of individuals, not their proportions in your calculations. Recognize also the modified degrees of freedom arising from use of only two variables, p and $q,$ in generating the three expected genotypic classes. Insofar as only one of these two variables is free to vary, you will be confined to one degree of freedom in the assessment of the P value.

PROBLEMS

1. In a randomly mating population of moths, 82% of the moths had pale wings; the remainder had dark wings. What does this say about whether the pale-wing character is dominant or recessive?

2. An artificial population is initiated with 50% AA individuals and 50% aa individuals. Calculate the allelic and genotypic frequencies after one generation of random mating, and demonstrate that Hardy-Weinberg equilibrium is reached after one generation of random mating.

3. In a population known to be in Hardy-Weinberg equilibrium, 75% of the population showed the dominant trait for the A gene locus. Determine the frequency of the dominant and recessive alleles.

4.   The following genotypic frequencies were found in a population: f(*AA*) = 0.05, f(*Aa*) = 0.20, f(*aa*) = 0.75. Is it possible to determine whether the population is in Hardy-Weinberg equilibrium?

5.   Using the information on cattle-coat color given in this chapter, calculate the expected proportions and numbers of cattle possessing white, roan, and red coat colors. Assume that cattle mate at random with respect to coat color.

6.   Using the expected numbers of cattle calculated for problem 5, compare the expected numbers with numbers observed (given in text), and determine whether the population is in Hardy-Weinberg equilibrium.

7.   The frequencies of genotypes in a population were as follows: f(*BB*) = 9/16; f(*Bb*) = 6/16; f(*bb*) = 1/16. These numbers represent the expanded binomial $(3/4 + 1/4)^2$. On this basis would it be certain that this population is in Hardy-Weinberg equilibrium?

8.   Tay-Sachs disease is an autosomal recessive disorder that causes severe mental retardation and other neurological disorders. Although rare, Tay-Sachs disease is most prevalent among Jewish populations of Eastern European descent. A survey found that 0.0016% of North American Jewish children were born with this condition.

        (i) What is the probable frequency of the Tay-Sachs allele in the population that was surveyed? (ii) What is the frequency of the wild-type allele for the locus? (iii) What is the frequency of carriers (heterozygotes) of the recessive allele?

9.   A sample of fifty individual plants from a population of the annual, *Impatiens biflora,* yielded the following genotypic data for the F and S allozymes of the phosphoglucose isomerase locus:

        30 *FF*   6 *FS*   14 *SS*

        (i) What are the genotype frequencies in the population? (ii) What are the frequencies of the fast (*F*) and slow (*S*) alleles in this population? (iii) What would the genotype frequencies be after one generation of random mating? (iv) Assume that any deviation from Hardy-Weinberg equilibrium is due solely to inbreeding. Calculate the inbreeding coefficient for this population.

10.   The genetic locus that determines the ability to distinguish the colors red and green is on the human X chromosome. The frequency of red-green color blindness, caused by a recessive allele at this locus, is 1.2% among males of a sample population. Assuming a Hardy-Weinberg equilibrium for the two alleles of this gene, (i) What is the frequency of the color-blind allele in this population? (ii) What is the frequency of female carriers in the population? (iii) What is the frequency of color-blind females?

11.   The spread of human settlement in tropical areas of Central and South America is causing available habitat for many plant and animal species to become fragmented. What evolutionary forces are important when a single population is divided into several smaller populations? How are genotype frequencies affected? Does the probability of mating among relatives increase or decrease? What is the consequence at loci having rare recessive alleles?

12.   A population of maple trees harbors a lethal recessive mutation that allows seeds to germinate, but the plants produce no chlorophyll and die shortly after germination. The allele frequency for this condition in the population is 0.01. (i) If a randomly mating population produces one million seeds, how many plants will lack chlorophyll? (ii) Will natural selection eliminate the allele after one generation? Why or why not?

13.   Consider the following data relating to two fitness components in *Drosophila melanogaster.* The numbers refer to the value *w*.

Genotype	aa	Aa	AA
Male virility	0.047	1.0	0.247
Viability	0.856	1.0	0.852

(i) How would you describe the fitnesses of the genotypes relative to one another? (ii) How do the fitnesses affect allele frequencies from one generation to the next? (iii) How do the fitness relationships affect the genotype frequencies observed at different points in the life cycle?

14. Many plants of the genus *Viola* (violets) produce two types of flowers on each individual plant: cleistogamous flowers that never open and are obligately self-pollinating; and chasmogamous flowers that open and are potentially cross-pollinated by other individuals. (i) What are the potential advantages and disadvantages of each pollination system? (ii) What would be the genetic consequences for a population that had pollination success only in the chasmogamous flowers? A population that had pollination success only in the cleistogamous flowers?

15. Samples of salmon from two rivers yielded the following data for polymorphisms for restriction fragment length (3.2, 3.8, and 4.0 kilobases [kb]) at the gene encoding the enzyme alcohol dehydrogenase.

Population I:

3.2 kb homozygotes	49
4.0 kb homozygotes	9
3.2/4.0 kb heterozygotes	42

Population II:

3.2 kb homozygotes	4
3.8 kb homozygotes	64
3.2/3.8 kb heterozygotes	32

(i) What are the allele frequencies in each population? (ii) Do the populations seem to be at Hardy-Weinberg equilibrium? (iii) What effect would migration of salmon from population II into the river where population I normally breeds have on the allele frequencies in the progeny of population I?

16. *Drosophila melanogaster* has four pairs of chromosomes. There is no recombination between any pair of homologous chromosomes in males of this species. How many genetically different types of gametes does a male produce? Females display recombination between members of all homologous chromosome pairs in meiosis. How many types of gametes does a female produce relative to a male?

17. Most new mutations, in all organisms, are deleterious. How, then, can mutation be the ultimate source of genetic variability?

18. In the peppered moth, *Biston betularia,* a single genetic locus affects the predominant coloration of the insect. The dominant allele *M* determines a dark gray pigment. A recessive allele, *m,* confers a light gray color on homozygous moths. A major source of mortality in peppered moth populations is predation by birds that eat moths resting on rocks or tree trunks having a color contrasting with that of the moth. How would you design an experiment to determine the relative fitnesses of the moths in different environments? Do you expect the fitnesses of each genotype to be the same in different environments? Is it possible, over many generations, to eliminate either allele in a given experimental population?

19. A population has the following allele frequencies at the ABO blood group locus: I^A, 0.152; I^B, 0.129; and I^O, 0.719. The I^A and I^B alleles are codominant with respect to one another, and both alleles are dominant over the I^O allele. If you sample one hundred persons from this randomly mating population, how many people would fall into each phenotypic category (blood types A, B, AB, and O), assuming a Hardy-Weinberg equilibrium?

20. The following autosomal RFLP genotype frequencies were found in a single sample of people:

Gene I:

3.6 kb homozygotes	0.16
2.8 kb homozygotes	0.36
2.8/3.6 kb heterozygotes	0.48

Gene II:

4.0 kb homozygotes	0.32
3.3 kb homozygotes	0.44
4.0/3.3 kb heterozygotes	0.24

(i) What are the allele frequencies at each locus? (ii) What would the genotype frequencies at each locus be at Hardy-Weinberg equilibrium? (iii) What are some of the reasons why one locus can be at Hardy-Weinberg equilibrium, whereas an independent locus in the same population may not?

Chapter 13

QUANTITATIVE GENETICS

Often traits show no simple correspondence between alleles of a single locus and the phenotype of the organism. Traits of this kind must be analyzed with different approaches from those described in preceding chapters. The nature, the causes, and the analysis of quantitative variation are explored here.

A. The Nature of Quantitative Variation

Many traits are continuous, showing no discrete categories from one phenotypic extreme to the other. Examples of continuously varying traits are height in human beings and yield in agriculturally important plants and animals. The number of possible phenotypes in a continuously varying trait is limited only by how finely one can measure differences. In contrast, meristic traits are those traits measured by counting, such as the number of seeds produced in fruits or eggs laid by chickens. There will always be a discrete value for a meristic trait. Finally, threshold traits have only a few phenotypic classes, even though numerous genes may determine the trait. As the number of the alleles of genes promoting a trait increases, the greater the likelihood an individual will express the trait. The point at which actual expression of the trait occurs is the threshold for the trait, a phenomenon underlying **penetrance,** the number of individuals of a population that express any manifestation of the trait. The expression of an abnormal number of toes in guinea pigs is an example of a threshold trait.

Quantitative variation is measured in populations by counting the number of individuals with a particular value for a trait. Depending upon the nature of the trait, and the number of individuals measured, the resulting distribution of values will be a smooth curve (if the variation is continuous or meristic and the sample size is large) or it will fall into a few discrete categories (for a threshold trait). The mean and variance are used to describe the variation of a trait. The mean, $\bar{x}$, is defined as:

$$\bar{x} = \Sigma x/n$$

in which x is the value of a trait for each individual that is measured, and n is the total population size. (The capital sigma, Σ, means "sum of.") The variance in a population is defined as:

$$s^2 = \Sigma(x - \bar{x})^2 / (N - 1)$$

The value $(x - \bar{x})^2$ represents the square of the deviation of each individual in the population from the mean for the population. Variance is a measure of the spread of the variation around the mean, and it is intuitively obvious that the larger the sum of the squared deviations from the mean, the greater the variance. Variance is related to sample size as well: for a given amount of variation around the mean, larger sample sizes (N) will have smaller variances.

B. Causes of Variation

Quantitative phenotypic variation results from the combined effects of genetic and environmental variation. Depending upon the trait, genetic or environmental effects may predominate, but in most cases neither factor governs variation exclusively. For some organisms, experiments can discriminate between potential sources of variation. For example, a plant breeder can determine the genetic basis of plant height by growing plants in a uniform environment. Any height differences are likely to reflect variation in genes controlling height, rather than environmental differences. Alternatively, growing a single strain of a plant in several different environments allows one to specify the effect of environmental variables on height.

These experimental approaches can be combined into a single experiment in which several different plant strains are grown in a variety of environments. If the strains respond differentially to the environments, there is a **genotype-environment** interaction. This is illustrated by Green Revolution crop strains. Compared with older varieties, the Green Revolution strains respond more favorably to applications of fertilizer (Fig. 13–1). The more rapid increase of yield with more fertilizer by one of the strains is an example of a genotype-environment interaction. If the lines were parallel, there would be no evidence for an interaction. In such experiments, it is essential that for each level of fertilizer, conditions are similar in other respects, so that the variation of fertilizer is the only independent variable. Otherwise, uncontrolled genotype-environment associations might make the experiment difficult or impossible to interpret.

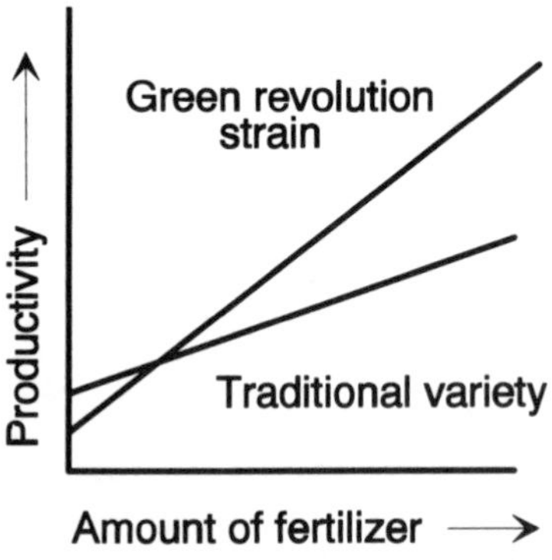

Figure 13–1 Relation of productivity to fertilizer applied to traditional and Green Revolution crops.

C. The Analysis of Genetic Variation

Environmentally induced variation is described as **environmental variance,** symbolized as σ_e^2. Conversely, genetically induced variation, or **genotypic variance,** is symbolized σ_g^2. The total variance in the population is defined as:

$$\sigma_t^2 = \sigma_g^2 + \sigma_e^2$$

The subscripts denote the total (t), genotypic (g), and environmental (e) components. The equation assumes that there are no interactions or associations between environmental and genotypic variance.

Environmental variance must be estimated before the genetic basis of a trait can be measured. For organisms that can be raised under controlled experimental conditions, the phenotypic variance in a genetically uniform strain grown in a single environment can be attributed to environmental effects. To obtain a genetically uniform strain, two inbred lines are crossed to produce an F_1 generation. The inbred parental lines are likely to be homozygous for all relevant genes, which means that all F_1 individuals will have the same genotype. The phenotypic variation among these F_1 plants is estimated as σ_e^2. The F_2 generation, however, will segregate for all genes that are heterozygous, and therefore will express both environmental and genotypic variation. Genotypic variance is calculated simply by subtracting the environmental component, estimated with the F_1 generation, from σ_t^2, the total phenotypic variance, estimated from the F_2 generation.

The magnitude of genetically controlled variation relative to the total in a population is measured as H^2, the broad-sense heritability, as follows:

$$H^2 = \sigma_g^2/(\sigma_g^2 + \sigma_e^2)$$

High values for H^2 imply that the genetic component of total variance is high in the particular environment in which the variances were estimated. Knowledge of the broad-sense heritability is useful in breeding programs, because high values of this parameter for a trait give promise that breeding and selection programs might succeed in modifying the trait.

Artificial selection programs provide another means of determining the genetic basis of a trait (Fig. 13–2). Unless a trait is under genetic control, selection cannot alter its mean value in subsequent generations. The degree to which artificial selection can alter the mean of a trait is dependent on the narrow-sense heritability for the trait in a given population. Here, we consider the mean value of a trait in the population, M, the mean value for the portion of the population, M^*, used as parents for a set of progeny, and the mean value of the progeny, M', that arise from this set of parents. The response to selection, which is dependent on the difference between the population mean, the mean of the selected individuals, and the narrow-sense heritability, h^2, is calculated from the formula:

$$M' = M + h^2(M^* - M).$$

If h^2 is small, the effect of even intense selection (large difference between M and M^*) is minimal, because the trait has very little genetic basis. If the difference between the population mean, M, and the selected portion of the population, M^*, is minimal, then the response to selection is negligible, even when the trait has a high narrow-sense heritability.

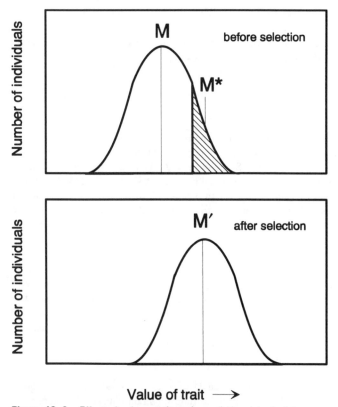

Figure 13–2   Effect of using a selected population (shaded) to produce offspring with a new mean value (M') of a continuous trait.

The above equation can be solved for h^2, the narrow-sense heritibility:

$$h^2 = [M' - M]/[M^* - M].$$

This rearrangement shows that narrow-sense heritability can be calculated as the response to selection divided by the intensity of selection on the population.

Narrow-sense heritability includes only additive effects of alleles, whereas the broad-sense heritability includes additive effects of alleles, dominance effects, and interaction among genes. The additive genetic variance (h^2) is important because it measures the responsiveness of the mean phenotypic value of a trait in a population to natural or artificial selection. Again, responses of traits to natural or artificial selection will depend on the magnitude of selection; that is, the difference between the selected and the total population ($M^* - M$), and the magnitude of additive genetic variance (h^2) for the trait in question.

There are limits to natural or artificial selection. As selected alleles increase in frequency and approach homozygosity, the response to selection will diminish. Another way of saying this is that the additive genetic variance for the trait diminishes; less variation remains after selection. Traits that have been under strong selection are thus expected to show little additive genetic variance. A second reason that selection cannot continue to produce change in a population is the occurrence of a correlated response to selection. Alleles under strong selection may have undesirable side effects that eventually resist further selection, despite the continued presence of genetic variation.

PROBLEMS

1. Calculate the mean, variance, and standard deviation for the following data on weight, in grams, of laboratory rats at 90 days of age.

   175, 202, 168, 154, 188, 195, 209, 191, 192, 177

2. It is difficult to measure any type of heritability without the ability to perform specific crosses. However, vast population studies enable estimation of heritability for some characters in human beings. Some estimated broad-sense heritabilities include: stature, 0.85; body weight, 0.62; systolic blood pressure, 0.57; diastolic blood pressure, 0.44; twinning, 0.50; and overall fertility, 0.1 to 0.2. Which of these characters is most likely to "run in families"? Which is least likely to "run in families"? If one of your parents and one of your grandparents has high blood pressure, should you be concerned about the likelihood of your having the same problem?

3. The narrow-sense heritability of percent protein in the milk of a particular cattle breed is estimated to be 0.55. A breeder has a herd whose percent milk protein varies from 4.0% to 25%, with a mean of 12%. The breeder inseminates only those cows whose percent milk protein is 18% with sperm from a bull whose mother and grandmother had 18% protein in their milk. What is the expected percent protein in the milk of the calves that result from these matings? How many generations of the same selection regime will it take to develop a herd that has milk with more than 17.5% milk protein?

4. The following data were collected from crosses between two inbred corn strains that differ in the percent oil in the kernels, as follows: Strain I: mean oil content = 4%, variance = 2; Strain II: mean oil content = 12%, variance = 3.

F₁ generation		F₂ generation	
Percent oil	Number of individuals	Percent oil	Number of individuals
2.5	0	2.5	10
3.5	20	3.5	60
4.5	40	4.5	80
5.5	90	5.5	100
6.5	120	6.5	120
7.5	150	7.5	100
8.5	220	8.5	160
9.5	190	9.5	90
10.5	100	10.5	110
11.5	50	11.5	100
12.5	20	12.5	50
13.5	0	13.5	20

(i) What is the environmental variance for oil content in this population? (ii) What is the genotypic variance for oil content? (iii) What is the broad-sense heritability for this character?

5.   Farmers and animal breeders have developed breeds of chickens that differ radically in egg weight, body weight, albumin content of eggs and time to sexual maturity. A trait that appears to be somewhat, but not wholly, resistant to selection for extreme values is the number of eggs per hen. What does this tell you about genetic variation for this trait in chickens? What does it tell you about the heritability of this trait?

6.   Two populations of the annual weed *Capsella bursa-pastoris* grow in adjacent fields. One field is mowed regularly; the other is never mowed. Seeds from the plants are taken to a greenhouse where they can be grown under uniform conditions, and the mean height of plants at time of flowering is measured for each population and for the F_1 and F_2 generations from crosses between representatives of each population. The data from these experimental populations was as follows:

Mowed field: mean = 6.2 cm, variance = 4.0

Not mowed: mean = 13.6 cm, variance = 10.2

F_1: mean = 10.2 cm, variance = 12.6

F_2: mean = 10.4 cm, variance = 20.3

How much of the observed variance in plant height at flowering is due to the environment in which the plants are grown? How much of the variance is due to genotype? What is the broad-sense heritability for this character?

7.   From each of the populations in problem 6, plants that flower at 8.0 cm were chosen for self-pollination or pollination by other members of the same population. When plants from the mowed field were crossed among themselves, the F_1 generation had an average height at flowering of 6.5 cm (variance = 8.2). When plants from the unmowed field were

crossed among themselves, the F_1 generation had an average height at flowering of 11.8 cm (variance = 16.6). What are the narrow-sense heritabilities for this character in these two populations? Suggest reasons for the difference in heritability between the two populations.

8. In the thousands of years that dogs have been domesticated, breeders have developed numerous breeds with many desirable traits. In many breeds of dogs, however, undesirable characteristics have appeared as well. The undesirable features include single gene traits such as retinal degeneration or deafness, and quantitative traits such as hypertension, skeletal abnormalities, or lack of normal intelligence. Why do these undesirable traits persist? Why do some traits get worse with more intensive breeding? Is it possible to continually select on the same character to create, for example, a twelve-foot dachshund?

9. The narrow sense heritabilities for several traits in domesticated cattle are: percent protein in milk, 0.54; feed efficiency, 0.34; milk yield, 0.30; calving interval, 0.01. Which of these traits would be most responsive to artificial selection? Which would be least responsive?

10. A plant breeder creates two separate strains of corn by intensive inbreeding for several generations. After developing the inbred strains, she crosses the two strains with each other to create a third strain. (i) What happens to the level of heterozygosity in the first two strains in the course of the inbreeding? (ii) How would you expect the performance of the inbred strains, in yield per acre, to compare with the original strains from which they were derived? (iii) How would you describe the third strain of corn? How would the performance of this strain compare with that of the inbred strains?

11. Several phenotypes of importance to human health, such as blood pressure, are influenced by many genes, most of which are not known. How can molecular genetics be combined with pedigree analysis and population studies to contribute to our understanding of the genetic basis of these quantitative traits?

12. A genetically uniform strain of wheat is planted in two different fields. The first field has had wheat grown on it for four consecutive years. The second field had soybeans on it for the two preceding years, and corn on it for the two years before that. The two fields were treated in an identical manner with respect to planting density, fertilization, water, and pest control. The wheat that was planted in the field where the crops had been rotated yielded nearly twice as much grain as did the wheat grown in the other field. Why would genetically identical organisms differ so radically in phenotype?

Chapter 14

EXTRANUCLEAR INHERITANCE

Extranuclear inheritance is often called **nonchromosomal, cytoplasmic,** or **non-Mendelian** inheritance. In this pattern of inheritance, the phenotype of offspring is determined by entities other than the chromosomal genes of the offspring. In most cases, the cytoplasm, donated by the egg, is the location of the inherited information or trait. The form of this information determines whether it is merely a transitory **maternal effect** or whether it is a case of persistent cytoplasmic inheritance. The DNA of organelles or symbionts underlies the latter category.

A. Maternal Effect

The development of an embryo is not determined by its genes alone. In higher plants and animals, the cytoplasm of the egg, which is a product of the maternal genotype, may also have a role for varying times after fertilization. This is because the activation of the genes of the zygotic nucleus is often delayed, and because materials contributed by the mother may override the activity of zygotic genes.

Maternal effect is transient; one or more chromosomal genes of the mother impart a phenotypic characteristic to eggs, and thereby to her offspring. However, the offspring cannot transmit the character to another generation if the offspring lack the appropriate genes. An example (Fig. 14–1) is the reddish larval skin and red-brown eye color of the moth *Ephestia*, in which a chromosomal gene *A* is required for the formation of kynurenin, a precursor of the red pigment. A mutation, *a*, blocks kynurenin synthesis. Matings of *Aa* females × *aa* males yield a genotypic ratio of 1 *Aa* : 1 *aa*, but phenotypically all progeny have red pigment. The reciprocal cross (*aa* females × *Aa* males), however, gives the expected result; namely, equal numbers of red and white larvae. The implication is that the *Aa* females can make eggs with enough kynurenin for the *aa* larva to make red pigment, even though these larvae cannot make kynurenin themselves. When adult *progeny* females of the first cross, all of which are phenotypically red, are mated individually with authentic white (*aa*) males, half the crosses (*Aa* × *aa*) yield the results of the first cross (all red), whereas the other half (*aa* × *aa*) yield all white progeny. In the second case, the genotypes of the female parents were actually *aa,* even though they had a red larval phenotype. Their capacity for making red pigment, which depends upon kynurenin, was exhausted, and they could not confer this character upon their own eggs.

A similar pattern of inheritance is seen in snails, in which the shells may coil to the right (clockwise, looking at the top) or to the left. The first cleavages following the fertilization of an egg determine this trait, and the cleavages are determined in turn by the genotype of the mother that lays them. In other words, cleavage direction is really an extension of the mother's genotype: mothers with the dominant allele for right-coiling will have right-coiled offspring, and mothers homozygous for the recessive allele, for left-coiling, will have left-coiled offspring, whatever the genotype of the offspring. This points up a difference from the example above of *Ephestia.* In snail-shell coiling, the mother's genotype overrides the genotype of the offspring; in *Ephestia,* a mother of the *aa* genotype (whatever her phenotype) cannot override the activity of the *A* gene in offspring.

Maternal effect is compared to conventional inheritance and to true cytoplasmic inheritance in Table 14–1.

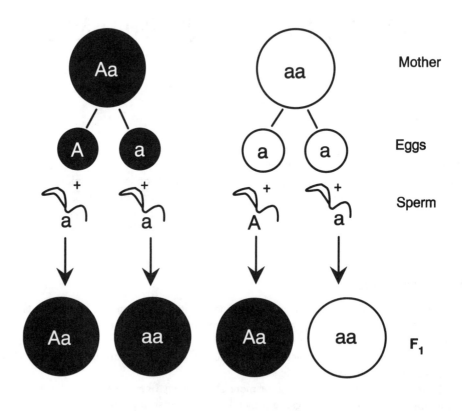

                    Aa female x *aa* male   *aa* female x *Aa* male

Figure 14–1   Reciprocal crosses of *Ephestia,* showing maternal inheritance of red color (dark). Crosses of *aa* females of the left cross to *aa* males will yield uniformly unpigmented progeny.

Table 14–1   Distinctions between conventional inheritance, maternal effect, and true cytoplasmic inheritance[a]

Type of inheritance	Source of genetic information	Examples
Conventional	Chromosomal genes of individual	Tall vs. short peas (Mendel)
Maternal effect	Chromosomal genes of mother	*Ephestia* larval skin and eye color; snail-shell coiling
Cytoplasmic inheritance[a]	DNA of individual's organelles or symbionts	*petite*-colony yeast (mitochondrion); variegated leaves of four-o'clock plants (chloroplast); killer factor (*kappa*) of *Paramecium* (symbiont)

[a] Maternal inheritance is one form of cytoplasmic inheritance; paternal inheritance of organelles is occasionally seen. "Cytoplasmic" inheritance is a better term for organisms in which mating types, rather than sperm and egg, determine fusion of gametes.

B. Cytoplasmic Inheritance

Non-chromosomal, replicating entities in the cytoplasm underlie "true" cytoplasmic inheritance. The criteria of this pattern of inheritance are (i) **persistence over many generations** and (ii) **a failure to segregate in a Mendelian fashion among progeny.** A corollary is that such traits cannot be mapped on a chromosome. In almost all cases, the genetic determinants of a cytoplasmically inherited trait are encoded in the DNAs of **organelles,** notably the chloroplast (chlDNA) or mitochondrion (mtDNA), or in the DNA of a **symbiont,** such as a virus or small intracellular bacterium. Because organelles and symbionts are usually transmitted only through eggs or ovules, but not by sperm or pollen, the term maternal inheritance is often used in place of the more general term, cytoplasmic inheritance. However, cases of plants in which pollen donates chloroplasts to offspring to the exclusion of the maternal contribution are known, and in some lower organisms, the gametes do not differ in their contributions of cytoplasm to the zygote.

Variegated leaves in four-o'clocks provide an example of organelle heredity, both at the vegetative and at the germ-cell level. The variegation of leaves is due to heterogeneity of chloroplasts, some carrying a normal gene and others carrying a defective gene for chloroplast function. Normally plant cells have a large number of chloroplast DNA molecules per chloroplast, and a large number of chloroplasts per cell. Once a mixed population of chloroplast DNA molecules has arisen by mutation, cell division will occasionally yield a cell without any normal chloroplasts, and all descendants of this cell will be white. The tissue of which this white lineage is a part will be variegated as long as both normal (with a normal or mixed chloroplast population) and white cells are present. However, some branches will arise from groups of cells that are pure white or pure green, and will remain purely white or green thereafter. This exemplifies random segregation of a cytoplasmic determinant in vegetative divisions.

Matings of flowers from white and green sectors of a plant will yield offspring that have the phenotype of the parent contributing the ovule; the pollen does not influence the chloroplast population of the offspring. Ovules made by variegated branches, however, may form green, white, or variegated offspring, depending upon the chloroplast DNA population in the ovule from which a particular plant arose. The progeny plants will continue the pattern of transmission in their offspring.

In yeast, the two haploid parental cells that act as gametes in matings are the same size. Here, the criteria for cytoplasmic inheritance (the term maternal is inapplicable) are nevertheless met if progeny do not segregate among meiotic products for a trait carried by one of the parents, and the determinant for the other trait is passed on indefinitely, both in mitosis and in crosses. In the analysis of organisms in which the gametic cells contribute equally to the zygotic cytoplasm, we must recognize the additional possibility that two different kinds of DNA might be present in the organelle population, and that pure sectors might occasionally arise by chance during mitosis, as they do in variegated four-o'clocks. This process is often referred to as segregation, but it is not equivalent to the Mendelian segregation of genes we see in meiosis.

The *petite* character (small colony) of yeast is an example of mitochondrial inheritance. A variety of *rho* mutations in mitochondrial DNA can abolish normal respiration, and this in turn leads to slow growth and very small colonies when the yeast are plated on an agar medium. In some crosses of *petite* × normal cells, all offspring are normal (*rho*$^+$), and the *petite* defect never appears in later generations (Fig. 14–2). This reflects the complete absence of mitochondrial DNA in some *petite* strains, and matings with a normal strain will restore normal mitochondria to all offspring. Other *petite* strains have a defective mitochondrial DNA, lacking segments of the normal mitochondrial genome, and in a cross to normal cells, the *petite* character infects all meiotic products of the cross. This reflects the greater rate of replication of the defective DNA than of the normal mitochondrial DNA. In these cases, the early products of cell division of meiotic products may initially be relatively normal, but all products yield small colonies in later

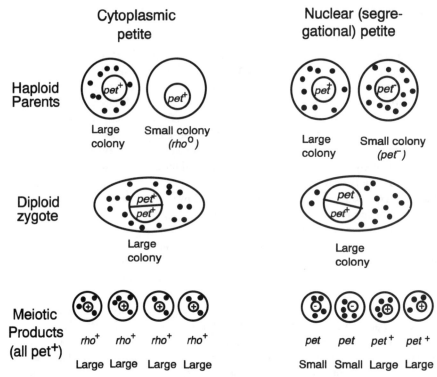

Figure 14–2 Inheritance of organellar (*rho°*, left) and nuclear (*pet⁻*, right) mutations of mitochondrial function in yeast. The carrier of the cytoplasmic determinant is mitochondrial DNA. In the case of cytoplasmic petites, the nuclear gene is normal, and the absence of the mitochondria determines the petite phenotype.

growth. Again, the criteria of cytoplasmic inheritance are met: persistence of the trait and lack of Mendelian segregation.

C. Effects of Chromosomal Genes

A complication of cytoplasmic inheritance is the influence of nuclear genes. The phenotype of a chromosomal mutation may closely resemble that of an organellar mutation. In other cases, a chromosomal gene may be required for the maintenance or expression of an organellar gene or mutation. Certain nuclear mutations of respiratory functions in yeast are examples of the first case. The ability to respire depends not only upon proteins encoded in mitochondrial DNA, but upon a much larger number of chromosomal genes as well. Therefore, "petite" yeast mutants have been isolated that, when crossed, yield a Mendelian ratio of 1 normal : 1 petite among haploid meiotic products. These are called **segregational petites** (gene symbol, *pet*) and behave like any other chromosomal mutant (Fig. 14–2).

Some chromosomal genes are required for the maintenance of a cytoplasmically inherited character. An example is found in the ciliate protozoan *Paramecium*, in which the nuclear *K* gene is required in its wild-type form to maintain a small intracellular bacterium called **kappa.** *Paramecia* with the kappa symbionts are called "killers" because they also excrete a toxin that kills other *Paramecia*. Without the dominant *K* gene, the symbionts are lost, and the cells simultaneously become sensitive to the toxin.

D. Organelles, Symbionts, and Their DNAs

As we saw above, the DNA of organelles is the location of genetic determinants for many cytoplasmically inherited traits. The DNA has a circular form, like that of bacteria (*see* Chapters 2 and 18), although it is considerably smaller. Organelles and symbionts impart a similar pattern of inheritance, due to the non-Mendelian distribution of their DNAs to the carrier's meiotic products. The similarity of inheritance patterns fits with the prevailing theory of the origin of chloroplasts and mitochondria. Organelles appear to have originated as symbionts, with photosynthetic bacteria evolving into chloroplasts, and nonphotosynthetic bacteria evolving into mitochondria.

What we now see as organelles are greatly reduced derivatives of prokaryotic cells, unable to survive on their own, but having all the molecular equipment needed for making a few of the proteins of the organelle and for independently replicating their DNA. The genes needed to make most proteins of the organelles, however, are located in the nucleus (hence the existence of segregational petites). The early evolution of the organelle probably included relocation, over a long period of time, of the original symbiont's genes into the nucleus of the host cell, and a coordination of functions of the nucleus and the evolving organelle. This process may have differed somewhat among organisms. For instance, one protein determined by mitochondrial DNA in the fungus *Neurospora* is actually encoded in the nucleus of another fungus, yeast. In general, however, mitochondrial DNA of most organisms has about the same number of different genes, the differences in size of the DNAs being accounted for by differences in non-coding DNA sequences, or by duplication of parts of the organellar genome.

Geneticists have actually been able to perform genetic experiments on organellar DNA in yeast and in the single-celled alga, *Chlamydomonas*. Matings of these species, in which cells of similar size fuse, bring two kinds of organelles into the same cell. This cell, a zygote, undergoes meiosis and in the vegetative divisions following meiosis, the random segregation of traits by which the organelles differ may be observed. By this means, it was discovered that organelles fuse, and organellar DNAs recombine sufficiently often that linkage maps of organellar DNA can be derived. The current method of establishing the nature and arrangement of organellar genes, however, has taken advantage of gene cloning and sequencing by which the structure of the organellar DNA may be studied more directly.

A final interesting attribute of organellar DNA in higher organisms, where it is transmitted by only one parent, is that little or no recombination among these hereditary units is possible. Therefore, the genetic differences in mitochondrial DNA among different populations of higher organisms arises almost solely through accumulation of mutations. This makes the nucleotide sequence of mitochondrial DNA favorable for the study of the evolution of higher organisms, including humans.

REVIEW QUESTIONS

1. What are the criteria for (i) maternal effect; (ii) maternal inheritance; (iii) sex linkage?

2. Stocks of the moth *Ephestia* were crossed; one stock (*AA*) had red larvae and the other (*aa*) had white larvae. All the F_1 progeny of the cross were red as larvae, and so were the F_2 moths, made by intercrossing the F_1s. What would you expect of testcrosses of individual F_1 and F_2 moths, in which (i) the tester was male and (ii) the tester was female?

3. The petite-colony character in yeast can be determined by mutations in mitochondrial DNA or in nuclear genes called *pet* genes. Two petite yeast strains were studied. Strain 1 was mated with wild type. The diploid was normal, and the meiotic products of this diploid segregated 1 normal : 1 petite. Strain 2, when mated with wild type, yielded a normal diploid. The diploid yielded only normal meiotic products. In a cross of strain 1 × strain 2, what would you expect the diploid and its meiotic products to be?

4. A strain of fruit fly with distinctively small adults was found in an isolated population. When mated with the normal strain, it yielded F_1 progeny that were all normal when the

female parent was normal and all small when the female parent was small. What is the most efficient way in which to test whether the small trait is due to maternal effect determined by a single pair of alleles or to true cytoplasmic inheritance?

5.   How would you characterize the difference between symbionts and organelles?

6.   A yeast strain with characteristically slow growth was found, and stocks of both mating types were obtained. Similarly, a strain with normal growth was isolated from an exotic source, and this too was represented by the two mating types. Intermatings of the slow-growing strains always gave slow-growing progeny, and intermating between the normal stocks yielded only normal progeny. However, matings of the slow and normal strains yielded diploid cells that were slow-growing. When induced to undergo meiosis, these diploids gave rise to uniformly slow-growing meiotic products, half of which were unable to grow without the addition of uracil to the medium.

Rationalize these data.

Chapter 15

MESSENGER RNA AND PROTEIN SYNTHESIS

Gene expression proceeds in two steps: transcription, the use of DNA in the synthesis of messenger RNA (mRNA), followed by translation, the use of mRNA as a code in the synthesis of polypeptides. Transcription occurs in the nucleus and is catalyzed by RNA polymerase. After export to the cytoplasm, mRNA is engaged by ribosomes, on which it is translated into a polypeptide. The molecular steps of translation are among the most complicated of cellular processes.

A. The "Central Dogma"

In Chapter 6, we saw that proteins catalyze almost all biochemical activities in cells. The amino acid sequences of polypeptides that make up proteins are encoded in the nucleotide sequences of corresponding genes. This "one gene-one polypeptide" concept leads us now to the details of how information in DNA is finally expressed in the form of protein amino acid sequences. The process is often referred to as the "Central Dogma" of molecular biology. The idea is that DNA transfers its information to RNA, and RNA in turn transfers its information to protein (Fig. 15–1). There are therefore two steps in the overall process: **transcription,** the synthesis of messenger RNA (mRNA), and **translation,** the synthesis of polypeptides. By using these words, we emphasize the flow of information, rather than a chemical transformation of one macromolecule into another. Since its formulation, the Central Dogma has been extended by the discovery of RNA viruses that transcribe their

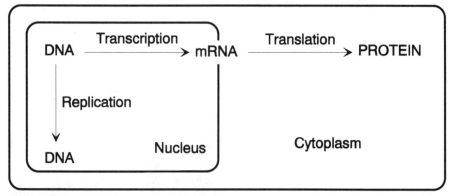

Figure 15–1   The "Central Dogma" of molecular biology: information flow in gene expression. The reverse-transcription process by which RNA viruses transcribe their RNA into DNA is not shown in the figure.

RNA nucleotide sequences into DNA. This "reverse transcript" can often integrate into chromosomal DNA.

B. Transcription

1. RNA structure

RNA differs from DNA in three ways: (i) it is single-stranded; (ii) it contains the sugar ribose in place of deoxyribose; and (iii) it contains the pyrimidine **uracil** (U) in place of thymine (T). Uracil and thymine have similar structures, and both make two hydrogen bonds with adenine. The nitrogenous bases A, G, and C are common to DNA and RNA. Otherwise, RNAs are nucleotide chains and, like DNA, carry information in their ribonucleotide sequences.

2. Transcription

The cell makes RNA nucleotide chains by using one of the two strands of DNA as a template (Fig. 15–2 and Table 15–1). RNA polymerase (RNAP), an enzyme in the nucleus of

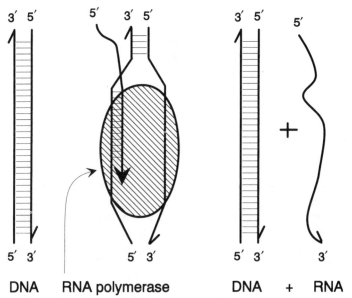

Figure 15–2   Steps in the synthesis of RNA by RNA polymerase (dark oval). RNA polymerase engages DNA at a promoter sequence, which identifies the strand and starting point for transcription. The strand chosen as the template has the 3′-to-5′ polarity; the transcript is made from its 5′ to its 3′ end.

Table 15–1   Summary of the components of the transcriptional and translational processes

Process	Component	Role
Transcription (RNA synthesis)	Template DNA	Code for RNA nucleotide order
	RNA polymerase	Polymerizes ribonucleotides
	Ribonucleotide triphosphates (rA, rG, rC, rU)	Substrates for RNA polymerase; building blocks for RNA
Translation (polypeptide synthesis)	mRNA (template)	Code for amino acids of polypeptide
	Ribosomes	Site of peptide-bond formation with sequential movement, 5′ to 3′, along the mRNA
	Aminoacyl-tRNAs	Transfer amino acids to growing polypeptide on surface of ribosome

eukaryotic cells, catalyzes RNA formation. At one end of a gene, a short nucleotide sequence called a **promoter** defines the point at which RNAP will bind, "melting" the DNA strands apart. This step identifies the DNA strand to be used, the 3'–5' strand, and the site at which transcription begins, lying a short distance down (in the direction of RNAP movement) from the initial binding site. Transcription begins at the 5' end of the new RNA chain. RNAP begins the synthesis of the mRNA chain, called a transcript, without a primer and polymerizes ribonucleotides into a linear chain, using ribonucleotide triphosphates. The chain is elongated continuously, with the same base-pairing rules that prevail in DNA. Here, rU pairs with dA, rA with dT, rG with dC, and rC with dG, in which **r** and **d** designate ribo- and deoxyribonucleotides, respectively. As RNAP moves, the DNA strands come back together behind it, forcing the new RNA molecule off the template. At the end of the gene, a **transcriptional termination site** defines the point at which RNAP leaves the DNA and releases the full-length mRNA molecule. The relationship between the DNA and RNA nucleotide sequences is shown in Figure 6–3, Chapter 6.

The process of RNA synthesis is called transcription because only the chemical form of the information changes, whereas the rules of the alphabet are the same. It resembles copying a handwritten message at the computer: the same English words are found in both the original and the copy.

3. RNA processing

Many mRNA molecules must be processed by trimming their ends or by undergoing **splicing** reactions, in which segments called **introns** are removed from internal locations. RNA splicing is almost never seen in prokaryotes, but is very common in eukaryotes. The discovery of splicing in eukaryotes was entirely surprising, revealing that for many genes of most organisms, the sequences that appeared in mRNA molecules were discontinuous in the DNA. While the DNA is chemically continuous, the segments corresponding to mRNAs, called **exons,** are interrupted by introns (Fig. 15–3). Thus, the genes and their polypeptide products are not strictly colinear. Special enzymes recognize key nucleotide sequences of introns and remove them before mRNA leaves the nucleus. This "splicing" of mRNAs renders exons of transcripts into continuous, colinear codes for the corresponding polypeptides. The number of introns varies from none or 1 to over 10 in many genes of mammals, and their length may greatly exceed the sum of the lengths of the exons.

In addition to splicing reactions, eukaryotes add special ends to most mRNA molecules. At the 5' end a special nucleotide, 7-methylguanidine (m^7G), is attached through its phosphate to the 5' phosphates of the first nucleotide of the mRNA. This so-called **cap** is added after the transcript is begun. The cap identifies the mRNA molecule as a message that encodes a protein, and is required if it is to be read efficiently by the translational machinery (see below). At the 3' end of mRNAs, a sequence of adenine nucleotides called a

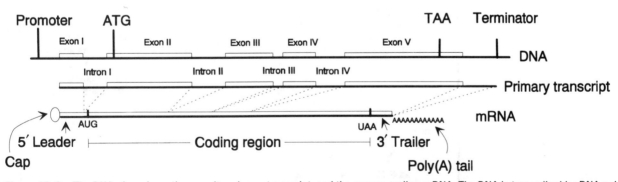

Figure 15–3 The DNA of a eukaryotic gene, its primary transcript, and the corresponding mRNA. The DNA is transcribed by RNA polymerase between a site to the right of the promoter (beginning of Exon I) and the terminator. The transcript is then spliced to remove introns, thereby bringing the exons (boxes) into continuity. The transcript is also capped with m^7G and "tailed" with polyadenylic acid (70–200 A nucleotides, depending upon the organism). The coding region of the mRNA lies between the first AUG and the first TAA that is in the same reading frame.

poly (A) tail, 50–200 bases long, is added by a separate enzymatic system after the mRNA is released from the DNA. The poly-A tail protects the mRNA molecule from enzymes that degrade RNA from the 3′ end.

4. The other kinds of RNA

Two other kinds of RNA, **transfer RNA (tRNA)** and **ribosomal RNA (rRNA)** are known. Unlike mRNA, they are never translated, but they perform essential functions in the translation process itself. These functions are described below and in Figure 15–4. Among cellular RNAs, however, the most diverse kind is mRNA, because mRNAs carry the information for all protein-coding genes to the cytoplasm where they are translated into polypeptides.

C. Translation

1. Components of the protein synthetic machinery

Protein synthesis depends upon three major components (Table 15–1). First, mRNA is the carrier of genetic information ("messages") from the nucleus. In eukaryotes, each mRNA molecule encodes only one polypeptide chain. (In prokaryotes, an mRNA molecule can encode as many as ten separate polypeptide chains, as described in detail in Chapter 19.) Second, ribosomes are particles consisting of ribosomal proteins and ribosomal RNAs. Ribosomes are the sites on which the polymerization of amino acids takes place under the direction of mRNA. Third, aminoacyl-tRNAs are short RNA molecules, called transfer RNAs (tRNAs), joined to specific amino acids. These molecules deliver amino acids to the ribosome and "read" the mRNA according to the genetic code for amino acids (Fig. 6–3, Chapter 6). The tRNAs are, in effect, adaptors that place amino acids in the order dictated by the mRNA.

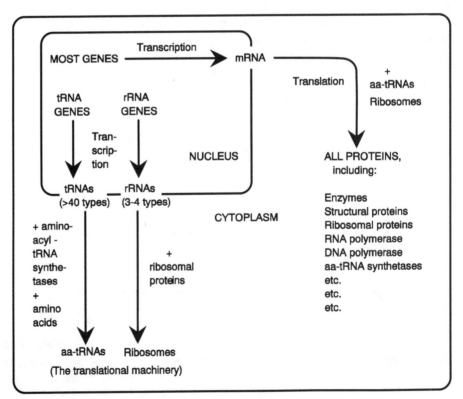

Figure 15–4   Origins, diversity, and roles of the various types of RNA (*see* text).

Ribosomes have large and small subunits, each one containing about equal amounts by weight of RNA and protein. The ribosome as a whole contains three (prokaryotes) or four (eukaryotes) different rRNA molecules. rRNAs are transcribed from three or four corresponding ribosomal RNA genes in the nucleus, but rRNAs are never translated. The four different rRNA molecules are part of the structure of the ribosome and participate in the chemical steps of protein synthesis. Ribosomal proteins assemble around rRNA molecules and with each other to form ribosomes. Ribosomes contain 54 polypeptides in prokaryotes, and 79 in eukaryotes. All ribosomes in the cytoplasm are the same. **Ribosomes do not dictate the sequence of amino acids in the polypeptides that they make; mRNA has that function.**

Aminoacyl-tRNAs have between 70 and 90 nucleotides and serve as adaptors for the amino acid attached at one end. Many of the nucleotides of each tRNA molecule bind to one another by internal base-pairing, which folds the tRNA into a specific, compact shape. The amino acid of an amino acyl-tRNA is attached by its carboxyl group to the 3′ end of the tRNA nucleotide chain, which lies at one end of the folded molecule (Fig. 15–5). An **anticodon** lies at an internal loop in the folded structure. The anticodon of a given tRNA is complementary in base sequence to at least one codon of mRNA. Certain anticodons can pair with more than one codon (always a synonymous one), because base-pairing rules are relaxed on the ribosome, and because tRNAs have modified bases that give them more flexibility in pairing. Thus, considerably fewer than 61 tRNAs can translate the 61 meaningful codons in mRNA. There are no tRNAs for nonsense codons.

Each amino acyl-tRNA has the proper ("cognate") amino acid attached to it, corresponding to its anticodon. The attachment of cognate amino acids is carried out by **amino acyl-tRNA synthetases,** a set of 20 or more enzymes responsible for the synthesis of the 20 types of amino acyl-tRNAs. Each synthetase must recognize (i) a single amino acid and (ii) the tRNAs that legitimately carry this amino acid. In this linking of tRNAs to cognate amino acids lies a chemically crucial step of translation, because the process brings the nucleotides of the triplet code and the 20 amino acids of polypeptides together in a single type of molecule. Amino acyl-tRNAs are made constantly by the aminoacyl-tRNA synthetases,

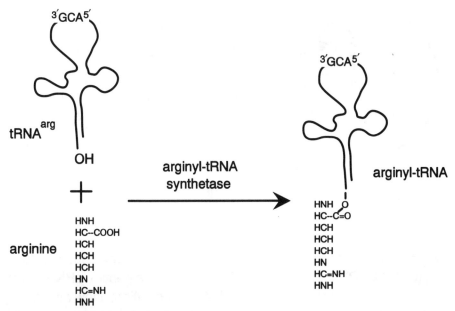

Figure 15–5 Amino acyl-tRNA formation, in which arginine is the amino acid. Note that the anticodon for arginine is 3′-GCA-5′, which pairs in an antiparallel fashion with one of the arginine codons (5′-CGU-3′).

keeping up with their use in protein synthesis (Fig. 15–4). A tRNA is recycled, after trans-ferring its amino acid to a growing polypeptide, by its attachment to another molecule of its cognate amino acid.

2. The process of translation

Translation takes place in the cytoplasm of eukaryotes, and begins with the small subunit of a ribosome binding to the 5′ cap of mRNA. The subunit acquires several proteins called **initiation factors** and moves to the first AUG triplet (for methionine, or Met) of the mRNA, where it binds to a molecule of methionyl-tRNA. This defines the starting point of the polypeptide chain and the reading frame used for translation (Fig. 6–3, Chapter 6). (In prokaryotes, the initiating amino acyl-tRNA is the modified formylmethionyl-tRNA). While methionine becomes the initial amino acid of the nascent polypeptide, this methio-nine is often removed after the polypeptide is completed. Only when the small subunit en-gages the initiating methionyl-tRNA does the ribosome fully assemble; the small subunit that began the process binds to a large subunit. The process of **polypeptide elongation** (Fig. 15–6) consists of several steps which repeat cyclically. A tRNA, its anticodon paired with a mRNA codon, lies in the left position of the ribosome. At the other end of the tRNA

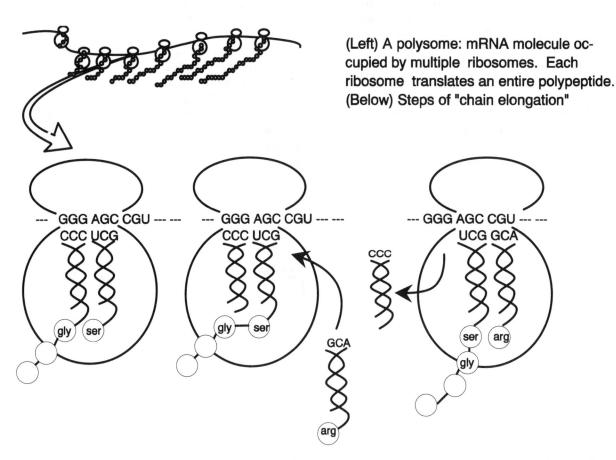

(Left) A polysome: mRNA molecule oc-cupied by multiple ribosomes. Each ribosome translates an entire polypeptide. (Below) Steps of "chain elongation"

1. Peptidyl-tRNA at left, amino acyl-tRNA at right.

2. Peptide bond formed (GLY - SER).

3. Ribosome moves one codon to right. tRNA evicted as peptidyl-tRNA moves to left-hand position. New amino acyl-tRNA enters right-hand position.

Figure 15–6 Protein synthesis.

is the corresponding amino acid, attached to all of the previously polymerized amino acids of the growing polypeptide chain. This tRNA, with its peptide, is called **peptidyl-tRNA.**

The right-hand position of the ribosome is available for occupancy by an amino acyl-tRNA. The amino acyl-tRNA that occupies it will be the one called for by the mRNA codon in that position, the three nucleotides following the codon at the left-hand position. After the new amino acyl-tRNA occupies the right-hand position, the bond between the left-hand tRNA and its attached peptide breaks, and a new bond forms between the peptide and the amino group of the new amino acyl-tRNA.

The ribosome then moves three nucleotides (one codon) down the mRNA. In the process, the tRNA in the left position, no longer attached to its peptide, is evicted from the ribosome as the tRNA in the right position (now a peptidyl-tRNA) "walks" into the left position. This leaves the right position open for another amino acyl-tRNA. The tRNA evicted from the ribosome is reused in the cytoplasm by the proper amino acyl-tRNA synthetase to regenerate a new molecule of amino acyl-tRNA.

Elongation of the polypeptide stops when the ribosome encounters one of the chain-terminating triplets, or nonsense codons, as stated earlier. At that point, the ribosomal subunits, mRNA, and the finished polypeptide dissociate from one another. The ribosomal subunits may then attach to another mRNA of any kind. mRNA molecules can be translated a number of times, and several ribosomes can translate an mRNA simultaneously. An mRNA with a number of ribosomes on it at one time is called a **polysome** (Fig. 15–6). Each ribosome must traverse the entire length of the polypeptide coding sequence to make a complete polypeptide. Most mRNAs are unstable, and break down after minutes to hours, depending on their nucleic acid sequences. The full significance of this will become clear in Chapter 19. Briefly, however, a short life of mRNA molecules makes it possible for the cell to change, in a rapid fashion, the proteins it makes by regulating only the rate of synthesis of mRNAs. mRNAs no longer appropriate at the time of the change will disappear quickly.

REVIEW QUESTIONS

1. Transcription and translation are template-directed processes. In each case, (i) what is the template; (ii) how is it "read"; (iii) what is the product; and (iv) how do the coding rules differ?

2. What are the main participants and their roles in transcription?

3. What is the role of amino acyl-tRNA synthetases in the process of translation?

4. What are the main participants and their roles in translation?

5. We speak of the "coding ratio" in gene expression. Which of the following statements best defines this phrase? (i) There are three synonymous codons for each amino acid; (ii) There are three amino acids corresponding to each codon; (iii) There are three bases in each codon specifying an amino acid; (iv) There are three tRNA's for each codon; (v) There are three times as many amino acids in proteins as there are nucleotides in DNA.

6. Describe the types and origins of RNAs participating in translation.

7. The synthesis of nucleic acids and protein is polarized, proceeding from one end to the other. What is the polarity of synthesis of (i) DNA nucleotide chains, (ii) RNA nucleotide chains, (iii) polypeptide chains, and (iv) the template DNA strand used in transcription?

8. How does an mRNA molecule differ from the primary transcript from which it is made?

Chapter 16

THE MUTATIONAL PROCESS

Mutations are heritable changes in DNA, ranging from substitutions of nucleotides to drastic changes in chromosome structure or number. Each category of mutation may be induced by way of damage to DNA or chromosomes. Cells have several mechanisms of repairing damage in DNA that counteract the effects of mutagens. However, even without mutagens, mutations appear spontaneously and are usually deleterious. Even rare beneficial mutations appear without any relation to the need of the organism. Mutational variation, arising randomly, is the raw material of evolutionary change.

Chapters 6 and 10 describe several types of genetic changes that can be called mutations. They are listed and defined in Table 6–2, Chapter 6. Here, we consider the ways in which mutations arise, either spontaneously (without specific induction) or through the action of DNA-damaging agents, and the ways in which the cell can intervene before damage becomes mutation.

A. Occurrence and Induction of Mutation

Base-pair substitutions arise by a variety of mechanisms. The simplest to understand intuitively is the occurrence of a base-pairing error during replication. The H-bonding properties of nucleotides change very occasionally owing to a rare shift of electrons. In this state, erroneous base-pairs may form and illegitimate bases may be incorporated into DNA. All DNA polymerases have an "editing" function which removes most erroneous bases immediately after they are inserted, before adding the next nucleotide. However, some are not removed, and this is the source of many spontaneous mutations—those arising with no obvious external induction. To visualize the process, consider replication of an A-T base pair which becomes an A-C base pair because C is mistakenly incorporated into one daughter molecule (Fig. 16–1). This is called a **heteroduplex**—a DNA with a mismatch. At the next replication, the strands separate, with A in one chain and C in the other. Synthesis of new chains yields an A-T (wild-type) pair in one DNA molecule and a G-C (mutant) pair in the other. The latter will be virtually immortal in further replication because it is chemically normal DNA.

Mutagens such as ultraviolet light (UV) induce base-pair substitutions through damage to DNA, followed by complex error-prone repair processes, discussed below. Chemical mutagens also induce base-pair substitutions. Some chemical mutagens are base analogs. These compounds resemble normal nitrogenous bases and are either incorporated into DNA during replication or interfere with DNA polymerase action, causing base-pairing errors. For instance, **5-bromouracil (5-BU)** is an analog of thymine and pairs with adenine. Brief exposure of cells to 5-BU leads to the synthesis of DNA with 5-BU in some positions. Its base-pairing fidelity is poor, however, so the 5-BU nucleotide, after becoming part of a nucleotide chain, can occasionally pair with guanine (G) at a later replication.

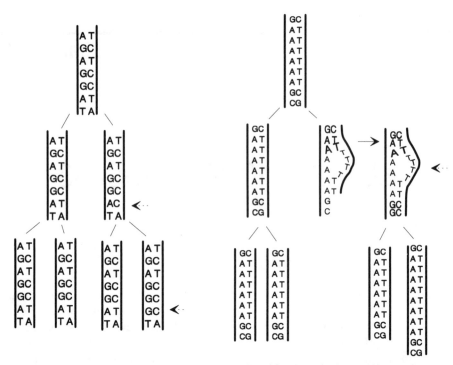

Figure 16–1 Origin of a base-pair substitution (AT to GC, left) and a frameshift mutation (addition of two T nucleotides, right) during DNA synthesis. In each case, a heteroduplex forms through an error in the first replication, and one derivative becomes a homoduplex mutant in the second division.

When a heteroduplex 5-BU/G mismatch replicates, the 5-BU on one chain will with high probability pair again with adenine, its usual partner, as a complementary chain forms. The guanine on the other strand pairs with cytosine, the two forming a mutant base pair. The net result, as in the example above, is that the original TA base pair has been replaced by a CG base pair in one of its descendants. The substitution will be permanent because it is normal DNA.

Other chemical mutagens, such as **alkylating agents,** nitrous acid, and nitrosoguanidine, cause base-pair substitutions by chemically altering normal bases. The altered bases may either impart poor fidelity during the next replication, leading eventually to base-pair substitution, or they may be repaired by error-prone processes similar to those seen after UV irradiation (*see* below).

Frameshift mutations may occur naturally by a slippage and faulty pairing of DNA chains such that one or more bases are left out or inserted during replication. These errors occur more frequently in runs of 5 or more A-T base pairs, which are more weakly hydrogen-bonded than G-C base pairs (Fig. 16–1). Frameshifts occur more frequently after treatment of replicating cells by chemicals, such as the acridines, known as **intercalating agents.** These compounds, as the name implies, slip between base pairs of DNA and cause errors during replication or repair.

Finally, chromosomal aberrations (called macrolesions; *see* Chapter 10) are induced by physical and chemical agents. X rays are very efficient at breaking chromosomes, and will often lead to deletions, translocations, and inversions. The same is true of UV, but the mechanism is much more indirect. Changes in chromosome number can be induced by special agents that interfere with the enzymes that normally ensure chromosome disjunction, or others, such as colchicine, that block the formation of the meiotic or mitotic spindle. If mitotic spindle fibers are not allowed to form, a nuclear membrane can reconstitute itself around the doubled chromosome set, thereby rendering the cell polyploid.

B. DNA Damage and DNA Repair

1. DNA damage

DNA damage and mutation are distinct phenomena. Damaged DNA is chemically abnormal DNA. Most DNA damage must be repaired, as we shall see below, if there is to be proper replication and function. Mutant DNA, on the other hand, has an altered, but chemically normal nucleotide sequence. Mutagens cause mutations by first damaging the DNA in some way. Many types of damage to DNA occur on only one strand, rendering the DNA heteroduplex. However, the undamaged complementary strand offers an opportunity for repair of the damaged strand by using the undamaged strand as a template for correction. Thus a heteroduplex DNA, once it arises, has three possible fates:

- failure to replicate, owing to severe and irreversible damage.

- replication, allowing at least the undamaged strand to serve as a template for normal replication. The damaged or altered strand, if it can serve as a template, may regenerate the heteroduplex (chemically modified base), or it might become a mutant homoduplex (if it originated as a nucleotide mismatch).

- repair, by which the damaged duplex is restored to the original or to a mutant homoduplex.

2. Repair of UV-induced damage

Ultraviolet light (UV) having a wavelength of approximately 260 nm is absorbed by the nitrogenous bases of DNA. The energy absorbed causes a characteristic chemical change, namely the formation of two covalent bonds between consecutive pyrimidines, usually two thymines, on one strand of DNA. The bonded pyrimidines constitute a pyrimidine dimer. A small dose of UV induces few dimers; heavy doses induce many more. Pyrimidine dimers are a serious form of damage, because the bonded pyrimidines cannot base-pair with purines, and replication using a dimer-containing strand as a template causes a gap in the new DNA strand. The gap in the new strand, when it separates from its complement at the next replication, leads to DNA breakage. In bacteria, this is a lethal event.

At least four major repair mechanisms that eliminate pyrimidine dimers or defer their effects on replication are found in all living things (Fig. 16–2). These mechanisms mitigate the effect of solar UV radiation to which most organisms are exposed. Some of the repair mechanisms are imperfect, however, so that errors in restoring the normal nucleotides lead to mutation. Thus UV light, besides killing cells, leaves an increased frequency of mutants among the survivors.

a. Photoreactivation.　This is a simple, error-free mechanism that breaks the bonds between pyrimidines and thereby restores the DNA to its undamaged state. It does not require synthesis of new DNA. The process is catalyzed by the **photoreactivating enzyme,** which recognizes pyrimidine dimers in DNA molecules. The enzyme requires activation by visible light (hence the term photoreactivation) in order to function.

b. Excision repair.　This mechanism is also error-free, at least in bacteria, where it has been studied extensively. The mechanism is implied by the name: an enzyme system that recognizes pyrimidine dimers cuts 20–30 nucleotides of the damaged chain, including the pyrimidine dimer, out of the DNA. A specialized DNA polymerase replaces these nucleotides, using the remaining chain as a template, followed by DNA ligase to make the repaired DNA chain continuous. The short segment of single-stranded DNA with the pyrimidine dimer is degraded.

c. Recombination repair.　When a pyrimidine dimer in a replicating DNA molecule enters a replication fork, where the nucleotide chains separate, a gap will appear in the new chain as it is formed opposite the dimer (Fig. 16–2). This is because pyrimidine dimers cannot direct the incorporation of purines into the new chain. This does not lead to breakage of the duplex, because the nucleotides on either side of the gap are hydrogen-bonded

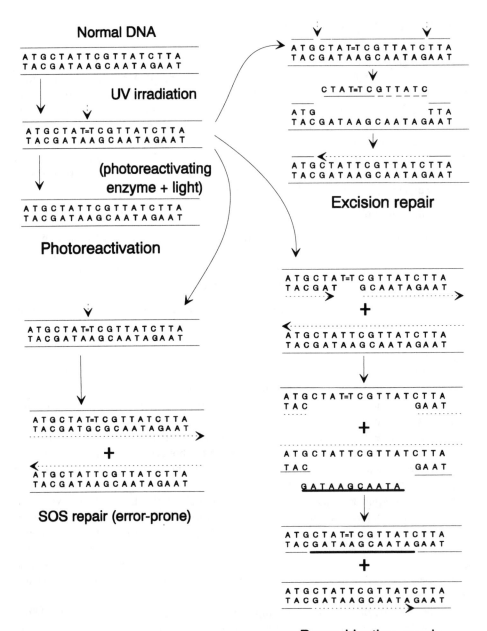

Figure 16–2   DNA repair mechanisms acting on pyrimidine dimers. The figure understates the number of nucleotides removed in excision repair; it is normally 20–30. SOS repair is depicted in terms of the products of the next replication. The bases opposite the pyrimidine dimer (which still remains) are random; G and C are shown. The strand containing these erroneous bases will direct the synthesis of a new strand with complementary, mutant bases at the following replication. Recombination repair is also depicted at a point following a replication, in which one product has a single-strand gap. The following events take place even before the two daughter molecules leave the vicinity of the replication fork. In these steps, the normal chromatid donates a segment of its lower chain as a replacement for the gapped strand of the dimer-containing duplex. DNA polymerase resynthesizes the missing segment of the lower chromatid.

with the template chain. As the resulting double-stranded molecule, with one gapped strand, leaves the fork, it is accompanied by a new, normal sister chromatid, a double-stranded DNA that is not damaged at the same point. The cell has an unusual mechanism by which it removes a segment of the new DNA strand with the gap in it and replaces it with a normal segment cut from the sister duplex. That is, it actually excises DNA from the sister and patches the gap. The immediate result is (i) a bigger gap in the sister duplex, and (ii) restoration of completely double-stranded DNA in the other duplex—which is heteroduplex because it retains the pyrimidine dimer.

Recombination repair continues by filling in the large gap in the sister duplex by DNA polymerization. The original pyrimidine dimer on the other chromatid may be repaired before the next replication. These complex maneuvers prevent DNA breakage when the strands of a DNA with a single-strand gap separate during replication. The recombination repair mechanism therefore does not really repair the pyrimidine dimer; only the gap caused by the dimer during replication is repaired. However, this gives time for other mechanisms to repair the dimer. All of the steps of recombination repair are error-free, although repair of the remaining pyrimidine dimer may involve error-prone processes (*see* below).

d. SOS repair. This mechanism is well defined in bacteria, but has counterparts in other organisms. The idea is implied by the SOS designation, taken from the maritime distress call ("Save Our Ship"). Cells heavily irradiated with UV become busy repairing DNA, a byproduct of which is cast-off, single-stranded DNA. Single-stranded DNA induces many new repair activities, including an error-prone DNA polymerase, to appear in the cell. The error-prone polymerase, when it encounters pyrimidine dimers in template DNA, inserts random—and thus in most cases erroneous—nucleotides in new strands when using the damaged chain as a template. Upon later replication or excision repair, the new, erroneous nucleotides are immortalized as erroneous, chemically normal nucleotide *pairs,* or mutations. SOS repair is probably the major source of mutations arising from UV irradiation.

3. Repair of other types of DNA damage

A variety of DNA-damaging chemicals, noted above, cause damage to the nucleotides of the duplex. Cells of all types can recognize these damaged nucleotides. Cells repair them by chemically reversing the chemical change, or by excising them in the manner described above for excision repair, restoring the DNA sequence to normal. One may ask what happens to a heteroduplex with a mismatch of normal nucleotides, (e.g., A-C). In that case, the repair enzymes involved have a choice, and because mutant base is not chemically recognized as erroneous, the repair systems excise the normal base in many cases, thereby creating a homoduplex mutant after filling in the gap. Thus mismatch repair involving normal bases is an inherently error-prone, mutagenic process.

4. Mutagens and cancer

Most cancer-causing chemicals (**carcinogens**) are mutagenic. This finding suggests that cancer originates in many cases by mutation. For instance, farmers and sailors frequently develop skin cancers on their hands and faces, exposed for years to sunlight containing substantial UV light. These cancers arise through the mutagenic consequences of pyrimidine dimers. A mutation of humans called *xeroderma pigmentosum* imparts such a sensitivity to sunlight that even moderate exposure is certain to lead to skin cancer in individuals homozygous for this gene. Significantly, the biochemical defects in such people are a deficiency for the photoreactivating enzyme or one of the components of the excision repair system.

More recent work has revealed that if certain proteins are inactivated by mutation, cells will begin to grow uncontrollably. Thus, over some time, the accumulation of mutations in the genes encoding these proteins will finally liberate a particular cell in a population to grow where such growth is normally suppressed. Mutations in these growth-controlling genes are therefore common to many cancers. This explains the carcinogenic character of most mutagens. Moreover, some people are born with mutations (in heterozygous or homozygous form) for one or more of these proteins. Such people are much more sensitive to carcinogens than others, because fewer mutation are required to render cells neoplastic. A predisposition to cancer is therefore seen in certain family lineages.

C. Spontaneous Mutation

1. Definition and character

Spontaneous mutations, by definition, have no known external cause. Their occurrence may be due to mutagenic conditions in the cell, such as free-radical formation; to an error of a DNA polymerase; or to unknown mutagens in the environment. Spontaneous mutations occur randomly: they may occur in any cell, at any time, and in any gene of a cell. However, the mutations do not occur with equal frequency in every gene or nucleotide; there are "hot-spots" in DNA that are particularly susceptible to mutation. An example is a run of A-T base pairs, a frequent location of frameshift mutations. However, there is no way of knowing which gene will become a mutant next, any more than one can predict which atom of a radioactive element will decay next.

In constant conditions, the mutation rate is constant, even if it is very low. The effect of a mutagen is simply to raise this rate. Because DNA of most genes is chemically similar, mutagens do not target specific genes. Spontaneous mutations occur at low frequencies, usually on the order of 1 in 10^6 to 1 in 10^8 per gene per generation. Because of the rarity of mutations, geneticists seeking mutants usually mutagenize populations of their experimental organism. This raises the frequency of mutants to levels at which they are easier to find.

2. Do mutations arise in response to need?

An important principle of mutation, and of spontaneous mutation in particular, is that mutations do not occur in response to the needs of the organism. They happen randomly, and have effects ranging from deleterious through indifferent to beneficial. This idea was hard for most people to believe, even after they accepted evolution of species from preexisting forms, because organisms display such precise adaptations to their environment. It was assumed by Lamarck, in the early nineteenth century, that evolution proceeded by the accumulation of variations (mutations) arising in response to need. Thus giraffes might have acquired long necks because they exercised them in obtaining leaves from high trees, and they passed the effect somehow—never adequately specified—to their descendants. A more modern example of beneficial mutation that Lamarck would have used to make his point is the appearance of antibiotic-resistant bacteria in environments containing antibiotics.

A more accurate picture has emerged more recently, based on rigorous experiments done by the geneticists Luria and Delbrück in the early 1940s. These experiments demonstrated that mutations arise randomly, even when they are not needed. Mutations are not often detected when they are not needed, because they occur so rarely. However, organisms that acquire certain mutations spontaneously are naturally selected in environments in which the mutation is beneficial. Organisms carrying such mutations grow or multiply more rapidly and become a prominent component of the population. The key point is that adaptation and evolutionary change are indeed directed by the environment, but not by molding individuals by special mutations. Instead, the environment directs adaptation by encouraging—selecting for—changes in the frequency of different alleles in large populations, in which rare beneficial mutations occur randomly all the time.

This view is less satisfying because it implies an enormous role of chance and "waste" on an enormous scale. Mutations are rare and most do not improve fitness. Therefore many mutations fail for each one that improves fitness, and huge numbers of animals and plants live with a "disadvantage" before advantageous alleles of various genes arise. However, this view is productive because it gives a reasonable and testable explanation of how organisms become so well adapted to their environments, some of these environments entirely novel in the course of geological history.

3. Proof that mutations preexist need

Although scientists were reasonably sure that mutations occur by chance during evolution, and not by necessity, they were nevertheless anxious to prove it rigorously. One such experiment was done by Joshua Lederberg during his seminal studies of bacterial genetics in the 1940s and 1950s. He used the bacterium *E. coli,* and asked whether streptomycin-resistance mutations would occur without exposing cells to streptomycin. He proved that they did so in an experimental regime called indirect selection, using a novel technique called **replica plating.**

In replica plating (Fig. 16–3), the face of a solid cylinder or disk, covered with velvet and secured by a ring, is pressed onto a Petri dish (called a "plate" by microbiologists) with bacterial colonies growing on agar medium. The nap of the velvet picks up a print of the colonies. The disk is then pressed onto a fresh agar plate, thereby printing the colonies of the original plate onto the new plate. The nap of the velvet simply acts as a forest of inoculating needles, faithfully transferring some bacteria from one plate to the other. This method allows microbiologists to print colonies from normal medium to a new medium containing, say, an antibiotic. The source of the antibiotic-resistant colonies is located on the original plate by the positions of those that grow on the new one.

Lederberg had to contend with the rarity of mutants (1 in 10^8 to 1 in 10^9 cells in the case of streptomycin resistance). He spread many bacteria over a plate of normal medium so that after it grew, the population formed an even lawn of cells. The population size may have been as great as 5×10^{10} cells. He then used a velvet replicator to transfer a sample of cells—perhaps as few as 1×10^6—to a Petri plate with medium containing streptomycin. Most of the transferred cells died, but two or three colonies grew in the presence of the antibiotic.

At this point the question can be posed rigorously: did the rare resistant cells that gave rise to the colonies acquire their resistance through their contact with streptomycin? Or were these cells present on the original plate that did not contain streptomycin? The answer was obtained by returning to the original plate with its lawn of cells and taking small samples at the points where resistant colonies appeared on the new plate (Fig. 16–3).

By spreading these small samples on a fresh plate not containing antibiotic, and then by replicating the resulting colonies to a streptomycin-containing plate, Lederberg found many more streptomycin resistant colonies than on his original replica. If he repeated this process, going each time to the positions on the antibiotic-free plate from which the resistant cells seemed to have come, he soon developed a pure, antibiotic-resistant strain (Fig. 16–3). The most important feature of this experiment is that antibiotic-resistant cells had arisen and had been purified without the slightest exposure to streptomycin. As a control, a series of transfers was made in the same way from positions on the streptomycin-free plate that did not give rise to streptomycin-resistant colonies. The descendants of these cells remained streptomycin-sensitive. The experiment as a whole proves that "beneficial" mutations can occur independently of an environmental challenge.

Many experiments done since those of Lederberg demonstrate that mutations occur all the time, if rarely, and that the environment does not induce specific mutations. The only sense in which the environment induces mutations is by raising the mutation rate of all genes of the organism, as, for example, in the presence of a mutagen. The environment merely selects—naturally, through differences in fitness—mutations that have already occurred.

No antibiotic   Streptomycin

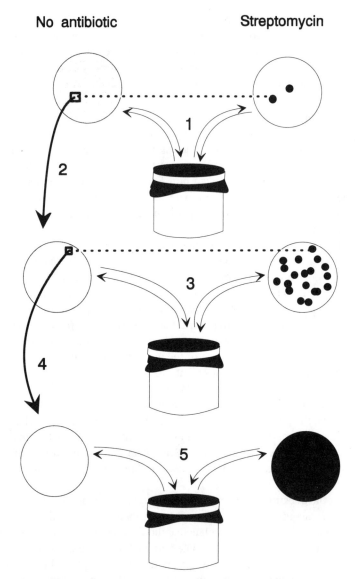

Figure 16–3   The experiment of Lederberg, demonstrating that mutations to streptomycin resistance occur prior to exposure to streptomycin. On the left, a lawn of cells grown on medium without streptomycin (grey) is replica-plated to a medium containing streptomycin
(step 1). The resistant colonies (black) are used to locate their origin
on the original plate. Bacteria from that point are spread on another
plate (step 2) lacking antibiotic. The process is repeated until a pure
culture is obtained (steps 3–5). The control, in which bacteria from
points on the antibiotic-free plate that did not yield resistant colonies
are followed similarly (*see* text), is not shown.

REVIEW QUESTIONS

1.   What are the major types of mutation? Characterize them briefly.

2.   How do mutation and DNA damage differ?

3. How does ultraviolet light (UV) cause mutation? Can the immediate effect of ultraviolet light be repaired before it causes mutation?

4. Can damage that does not involve the occurrence of chemically abnormal nucleotides be repaired? If so, how?

5. Some stocks of *Drosophila* or of *E. coli* have substantially higher mutation rates than normal, and the cause appears to be genetic. Can you think of the sorts of gene that might, when mutant, lead to such a phenotype?

6. A friend can't believe that adaptations of organisms to their environments occur wholly "by chance," as many people think scientists would have it. What would you do at least to clarify the question, if not to prove that chance plays a significant part in evolution?

Chapter 17

ALLELISM AND ANALYSIS OF COMPLEX FUNCTIONS

Geneticists use mutations to dissect complex cellular functions underlying viability, development, and behavior. The formal aspects of such analyses are presented here, using a simple biochemical pathway as an example. The chapter shows how simple genetic tests can provide a formal understanding of how genes interact in a multigenic phenotype.

A. A Biochemical Pathway

The sequence of enzymes that synthesizes the amino acid arginine in *Neurospora crassa* is diagrammed in Figure 17–1, showing all the intermediates in the process. It is linear for the most part, each enzyme modifying its substrate slightly as the starting material, glutamate, is transformed to the end product, arginine. A small tributary provides the intermediate carbamoyl phosphate, required for the conversion of ornithine to citrulline. The enzymes, represented by arrows, are numbered, and the genes encoding them are also given.

Chapter 6 explained how mutations can eliminate the function of a gene product such as an enzyme. In the early 1940s, Beadle and Tatum chose *Neurospora,* a simple fungus, to study mutations that affected the enzyme reactions catalyzing biochemical processes. Among their mutations were those that imposed an arginine requirement upon the organism. The mutants were of a class called **auxotrophs**—variants needing a nutritional supplement, in contrast to the original wild-type strain, a **prototroph.** The mutant and wild-type strains were distinguished by their ability to grow on minimal medium, a medium containing only the components essential to the growth of the wild-type strain.

In order to analyze this pathway thoroughly, the followers of Beadle and Tatum isolated a large number of different mutant strains that required arginine. They hoped that mutants lacking each specific function—that is, each enzyme—would be found among these strains. The collection of arginine-requiring mutants (generally known as *arg⁻* mutants) would reveal the effects of blocking each step of the pathway. Let us explore a hypothetical analysis of this sort.

B. Isolation of Mutants

Neurospora is haploid, but because it is a filamentous fungus, it has a number of nuclei coexisting in the same cytoplasm. The nuclei distribute themselves relatively evenly in the **hyphae** that make up the spreading tangle of filaments known as a **mycelium** (Fig. 17–2). *Neurospora* also makes aerial hyphae that bear asexual spores known as **conidia** (Fig. 17–2). Conidia have one to three nuclei. An investigator, using a suspension of conidia, can induce mutations in conidial nuclei with UV irradiation. The investigator then spreads the

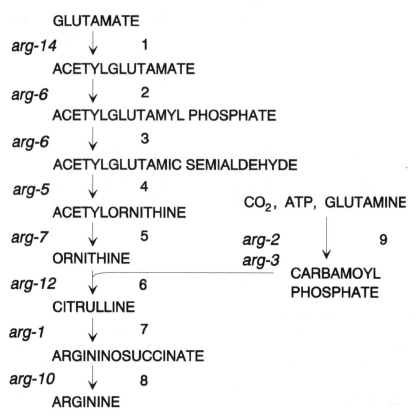

Figure 17–1 The arginine pathway of *Neurospora*, showing the enzymatic steps of synthesis (1–9) and the corresponding genes that encode the enzymes catalyzing the reactions.

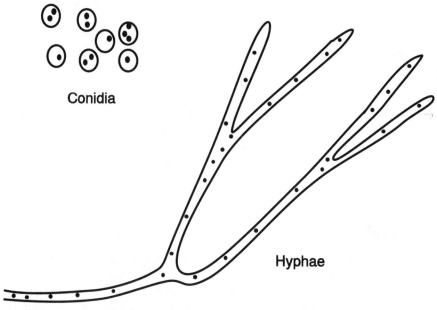

Figure 17–2 *Neurospora:* conidia and vegetative hyphae. The former, upon germination, grow into a branching set of hyphae that can grow in this manner indefinitely.

survivors on solid, arginine-containing medium, isolates and grows the resulting colonies, each arising from a single conidium, and tests them for their nutritional requirement.

Many tricks have been used to improve the efficiency of selecting mutants. In *Neurospora,* large numbers of *arg⁻* mutants may be obtained by first irradiating millions of conidia and allowing them to grow in a liquid, minimal medium. Only the wild-type conidia will grow, extending filamentous hyphae in all directions. After the culture has grown 12–18 hours, it is filtered through cheesecloth and allowed to grow some more. The filtration removes the growing, wild-type conidia, and the conidial population that remains becomes enriched for mutant, non-growing cells. After repeating this process several times, the remaining conidia are plated on agar medium, this one containing arginine, at a density that allows each conidium to form a separate colony. Mutant conidia that require arginine will grow, together with prototrophic conidia that have slipped through the filtration steps. There are of course many other mutants in the final population, representing potentially every other gene function in the cell. However, they will not grow on this medium, which has been chosen specifically for the isolation of *arg⁻* mutants.

The colonies growing on arginine-supplemented plates are isolated carefully to individual, small tubes containing the same medium, where they mature into full-grown cultures. These small cultures are tested on minimal medium to see whether they are auxotrophs. In our example, twenty–six *arg⁻* auxotrophic mutants, labelled a–z, were isolated.

C. Grouping the Mutants by Complementation

Even though we know the biochemistry of the arginine pathway, we must find a way of determining which mutants are alleles, and which are non-allelic. If we knew nothing of the pathway, we would be embarking on its analysis without even knowing how many genes to expect. Our question about allelism is really which mutations affect the same protein or polypeptide, and thereby simplify our large number of mutants into groups for easier study. We can later ask what mutational sites are affected by allelic mutations.

We begin with a test of allelism that sorts out the *arg⁻* mutants representing different genes. You might initially think that intercrosses would accomplish this: any pair of mutations that failed to recombine with one another to form wild-type progeny would presumably be allelic. Several problems make intercrosses a poor approach. One is that we must rely on negative evidence—a lack of recombinants. The second is that mutations may be in different genes that lie very close together on the chromosome, and only a large number of progeny will reveal recombinants in a mutant × mutant cross. The third is that even allelic mutations recombine at low frequencies if the mutations lie at different positions in the same gene (Fig. 17–3). Finally, all mutants will have the same mating type, which without intermediate crosses, makes them impossible to mate with one another.

In order to circumvent these problems, we use a **complementation test,** which tests the ability of the mutants to overcome one another's deficiencies functionally, rather than their ability to recombine genetically.

As noted above, *Neurospora* hyphae are actually filaments that contain many nuclei in a common cytoplasm. Because vegetative cells of *Neurospora* can fuse with one another, within a culture or between genetically different strains, the organism can form **heterokaryons,** cells with genetically different, haploid nuclei (Fig. 17–2). The formation of heterokaryons is automatic if the strains are closely related, as our mutant strains are. The fusions are not crosses, because (i) they involve fusions of vegetative structures, not sexual structures and gametic nuclei; and (ii) they happen only between cells of the same mating type, not different mating types. The heterokaryons thus formed are physiologically equivalent to heterozygotes of diploid organisms, because they contain entire genomes with different alleles of one or more genes in the same cell. It is not important that the alleles lie in different nuclei nor that the nuclei may depart from a strict 1:1 ratio. The important feature of the heterokaryon is the contribution, by the different nuclei, of mRNAs and proteins to a common cytoplasm. With that, dominance relationships between mutant and wild-type alleles of the same gene can be determined.

Consider one gene at a time. A heterokaryon formed between a wild-type strain and an arginine-requiring *arg-10* strain will grow because the enzymic deficiency of the mutant nucleus is overcome by the activity of the wild-type nucleus in the same cell. Therefore, we consider the wild-type *arg-10*⁺ allele dominant to the *arg-10* mutation. The same dominance relationships prevail between the wild-type and mutant alleles of any other of our *arg* genes.

The dominance relationship between a wild-type and a mutant allele is a functional relationship, not a positional one that we would assess by segregation and linkage analysis.

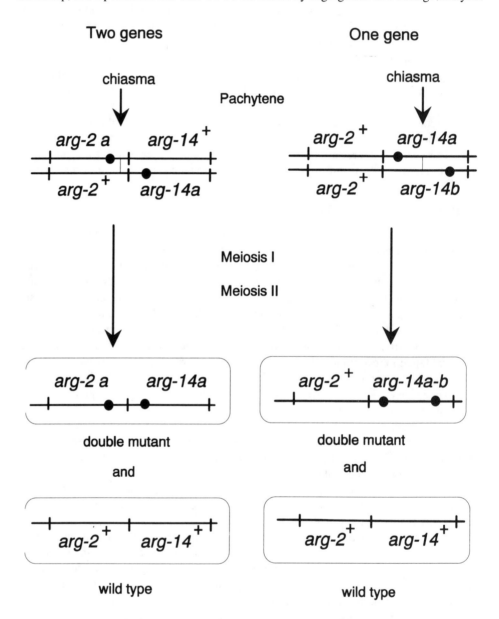

Haploid, recombinant meiotic products

Figure 17-3 Recombination during meiosis between two non-sister chromatids. The same segment of chromosome is shown in both cases: left, recombination between mutations in two different genes; right, intragenic recombination between two mutational sites in the *arg-14* gene. In both cases, owing to the proximity of mutational sites, recombination is extremely rare (<0.01%).

We can use the dominance relationships to determine which *arg⁻* mutations are allelic to one another, because two nuclei, if they carry mutant alleles of the same gene, cannot support the growth of a heterokaryon in minimal medium. Using symbols, if a heterokaryon is made with two mutations of the *arg-1* gene, *arg-1a* and *arg-1b,* neither nucleus has the genetic information for an active argininosuccinate synthetase protein (Fig. 17–4), and therefore neither nucleus can compensate for the other's deficiency. Thus, two mutations are allelic (= affect the same polypeptide) if a heterokaryon containing the two single-mutant nuclei fails to grow in minimal medium (Fig. 17–4). The heterokaryon is equivalent to a double-recessive homozygote of a diploid (e.g., *bw/bw* of *Drosophila*). This is one possible outcome of a **complementation test.**

If the two mutations in the heterokaryon damage different genes (e.g., *arg-1* and *arg-10*), then the heterokaryon will grow. The relevant genotype of the heterokaryon is [*arg-1⁺ arg-10* + *arg-1 arg-10⁺*], in which two simultaneous dominance relationships prevail (Fig. 17–4). Notice that the mutant nuclei have a complementary relationship: what one lacks in terms of function, the other can perform. This is the other outcome of a complementation test.

In practice, a complementation test of the *arg⁻* mutants is simple: the strains are inoculated together in all possible pairs in culture tubes of minimal medium. The inocula will fuse together, and the resulting heterokaryon will not grow if the mutations are allelic, and will grow if they are not allelic. An example of a complementation test is given in Figure 17–5, in which the grid represents all possible pairs of 26 mutants, with the **homokaryotic control** for each mutant on the diagonal. If you follow each row from the left to the diagonal, and then down, you will see that each mutant is paired with all others. With the + (growth) and 0 (no-growth) responses, strains carrying allelic mutations may quickly be determined. Notice that the results in this example are self-consistent: any two mutations that fail to complement (no growth) with a third mutant also fail to complement with one another. The 26 mutants can be grouped quickly into eight complementation groups, each containing mutants unable to complement with one another, but able to complement with mutants of all other groups.

In many systems, two mutants unable to complement with a third sometimes complement with one another. This is called intragenic complementation. It complicates, but does not invalidate the test as long as mutants unable to complement with all others of the group appear. The basis of intragenic complementation is an interaction of allelic polypeptides carrying different mutational lesions to form a multimeric protein with some function restored. Mutants carrying deletions and nonsense mutants are unable to contribute polypeptides to intragenic complementation interactions. These are therefore good testers to define—by lack of complementation—all other alleles of the gene.

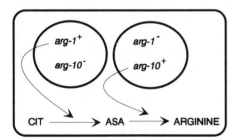

Growth on minimal medium = intergenic complementation.
DIAGNOSIS: MUTATIONS ARE PROBABLY NON-ALLELIC

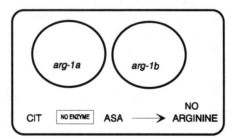

No growth on minimal medium = no complementation.
DIAGNOSIS: MUTATIONS ARE ALLELIC

Figure 17–4 Complementation tests for allelism, and the outcome in tests of non-allelic (left) and allelic (right) mutants.

```
        a
     a  0  b
     b  +  0  c
     c  +  +  0  d
     d  +  +  +  0  e
     e  +  +  +  +  0  f
     f  +  +  0  +  +  0  g
     g  +  +  0  +  +  0  0  h
     h  +  +  +  0  +  +  +  0  i
     i  +  +  0  +  +  0  0  +  0  j
     j  +  +  +  +  +  +  +  +  0  k
     k  +  +  0  +  +  0  0  +  0  +  0  l
     l  +  +  +  +  +  +  +  +  +  +  +  0  m
     m  +  +  +  +  0  +  +  +  +  +  +  +  0  n
     n  +  +  +  +  0  +  +  +  +  +  +  +  0  0  o
     o  +  +  +  +  +  +  +  +  +  0  +  +  +  +  0  p
     p  +  +  +  +  +  +  +  +  +  0  +  +  +  +  0  0  q
     q  +  +  0  +  +  0  0  +  0  +  0  +  +  +  +  +  0  r
     r  +  +  +  +  +  +  +  +  +  +  +  +  +  +  +  +  +  0  s
     s  +  +  +  0  +  +  +  +  +  +  +  +  0  0  +  +  +  +  0  t
     t  +  +  0  +  +  0  0  +  0  +  0  +  +  +  +  +  0  +  +  0  u
     u  +  +  0  +  +  0  0  +  0  +  0  +  +  +  +  +  0  +  +  0  0  v
     v  +  +  0  +  +  0  0  +  0  +  0  +  +  +  +  +  0  +  +  0  0  0  w
     w  +  +  +  0  +  +  +  +  +  +  +  0  0  +  +  +  +  0  +  +  +  0  x
     x  +  +  0  +  +  0  0  +  0  +  0  +  +  +  +  +  0  +  +  0  0  0  +  0  y
     y  +  +  +  0  +  +  +  0  +  +  +  +  +  +  +  +  +  +  +  +  +  +  +  +  0  z
     z  +  +  +  0  +  +  0  +  +  +  +  +  +  +  +  +  +  +  +  +  +  +  +  +  0  0
```

Complementation group	Mutants	Enzyme
I	*a*	3
II	*b*	7
III	*c f g i k q t u v x*	9
IV	*d h y z*	1
V	*e m n s w*	9
VI	*j o p*	4
VII	*l*	8
VIII	*r*	6

Figure 17–5 Complementation grid encompassing 26 *arg⁻* mutants, showing growth (+) or no growth (0) on minimal medium of all pairwise combinations. To the right of the grid, the mutants in each complementation group are shown (there is no intragenic complementation in this example), together with independent evidence (i.e., from direct tests) of the enzyme (*see* Figure 17–1) that is deficient in each case.

D. The *Cis-trans* Test: Meaning and Relation to Recombination

Complementation tests are also called ***cis-trans*** tests, and more explanation of their formalities is required. First, the term *cis-trans* refers to the relative positions of the mutations in the test. If the mutations being tested for allelism are located in different nuclei, they are said to be in ***trans.*** The term is identical to the term "repulsion" in the genotype *Ab/aB,* where the mutations of a double heterozygote lie on different homologs. **Complementation tests must be done in the *trans,* or repulsion, relationship.** The reason can be appreciated best if one considers the alternative, ***cis*** relationship (Fig. 17–6). The heterokaryon [*arg-1 arg-10* + *arg-1⁺ arg-10⁺*], where non-allelic mutations are in the *cis* relationship, would be normal, just as the *trans* heterokaryon was. (In all cases so far, we have assumed the mutations are recessive.) Now consider the heterokaryon [*arg-1a-b* + *arg-1⁺*], in which the two mutations of the *arg-1* gene lie in one nucleus (in fact, in one gene), paired with a nucleus carrying the wild-type *arg-1⁺* allele. Formally, the two *arg-1* mutations have been rearranged to reside in the same nucleus, just as we rearranged the *arg-1* and *arg-10* mutations to a *cis* relationship (Fig. 17–6). In doing so, we find that the [*arg-1a arg-1b* + *arg-1⁺*] heterokaryon can grow in minimal medium. This only proves that a wild-type allele is dominant to a damaged (in this case, a doubly damaged) one. Because both allelic and non-allelic mutations behave the same way in the *cis* arrangement, the *cis* arrangement is not useful in deciding allelism.

Non-allelic mutations

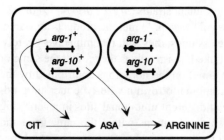

cis arrangement
Arginine is made:
demonstrates recessive-
ness of both mutations

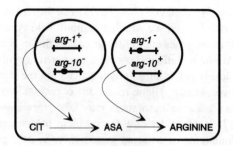

trans arrangement
Arginine is made:
demonstrates mutations
complement

Allelic mutations

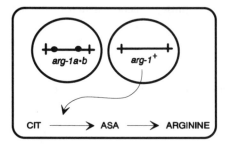

cis arrangement
Arginine is made:
demonstrates recessive-
ness of double-mutant gene

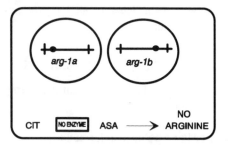

trans arrangement
Arginine is not made:
demonstrates mutations
fail to complement

Figure 17–6 The *cis-trans* test. Non-allelic (top) and allelic (bottom) are each shown in the *cis* and the *trans* arrangement, and the capabilities of each heterokaryon in making arginine is shown. The *trans* arrangement (taken from Fig. 17–4) is suitable as the test of allelism.

In order to resolve confusion, the meaning of the symbols used must be appreciated. The designations *arg-1⁻* and *arg-10⁻* refer in the discussion above to mutant **genes,** while the designations *arg-1a* and *arg-1b* refer to **mutations (or "mutational sites") within a gene.** As noted at the beginning of the chapter, genes are segments of DNA that must be intact in order to make a functional polypeptide. When "two mutations complement", it is really their wildtype alleles, one in one nucleus, the other in the other nucleus, that sustain the growth of the heterokaryon in minimal medium. Therefore, genes are the entities we refer to as alleles. Mutations (e.g., *arg-1a*) are usually less extensive than the gene in which they lie. When we find two mutations that fail to complement, we know by the mutant phenotype of the heterokaryon that the same gene is damaged in both nuclei, even though the gene in the two nuclei may be damaged at different points. One cannot restore function by putting two damaged alleles together in a common cytoplasm.

Taking this one step farther, let us see how allelic and non-allelic mutations **recombine** (Fig. 17–3). As you might expect, non-allelic mutations usually recombine detectably in meiosis. (Note that complementation tests are not done in meiosis, but in vegetative or somatic cells). If we were to test for the recombination of the tightly linked *arg-2⁻* and *arg-14⁻* mutations, we would mate the two single-mutant strains (*arg-2⁻ arg-14⁺* × *arg-2⁺ arg-14⁻*) and see whether haploid, wild-type (*arg-2⁺ arg-14⁺*) meiotic products appeared. If these genes are tightly linked, as this example assumes, they will recombine very rarely. Allelic mutations, however, are *always* tightly linked, inasmuch as they affect the same or closely neighboring positions (within the same gene) on the homologs contributed by the two parents. Allelic mutations occasionally recombine to form a wild-type meiotic product, because a crossover may take place between different mutational sites in a gene (Fig 17–3). Consider now the problem of distinguishing mutations within the same gene and mutations in adjacent genes: there is no rigorous distinction between their **recombinational** behavior, as Figure 17–3 demonstrates. That is why geneticists decide allelism by a functional complementation test, rather than a "positional," recombination test. The complementation test, fortunately, is much easier as well.

Finally, because complementation tests require two sets of genes in one cytoplasm, they may be done easily in diploids. We have seen that non-allelic, recessive mutations confer a wild-type phenotype on double heterozygotes (e.g., *eb⁺/eb bw/bw⁺*); the two mutations, *eb* and *bw,* complement. In order to obtain the diploid, a mating must be done between two parental strains, each carrying *one* of the two genes so that the *trans* arrangement is assured. Because of this, confusion often arises between the wild-type phenotype of the *diploid hybrid* containing complementing, non-allelic mutations and the wild-type phenotype of the *haploid meiotic products* arising through recombination of the same, non-allelic mutations.

E. Further Study of *Arg⁻* Mutants

The *arg⁻* mutants fell into eight complementation groups (Fig. 17–5). What else can we learn about them?

1. Growth on pathway intermediates

Many biochemical pathways have metabolic intermediates that can enter cells when added to the growth medium. These intermediates can be tested for their ability to support the growth of mutant strains. A test of the arginine mutants (Table 17–1) shows that many grow on citrulline as well as arginine, and others grow on arginine, citrulline or ornithine. No mutant able to grow on ornithine failed to grow on citrulline as well. These growth responses show that the order of the intermediates in the pathway was ornithine → citrulline → arginine, and that mutations create specific enzymatic blocks between citrulline and arginine, between ornithine and citrulline, or before ornithine. No more information can be obtained from nutritional studies, because other intermediates are too unstable or too impermeant to support growth. However, the information is so far consistent with the view that each mutation affects only one enzyme or protein, leaving others intact.

Table 17–1 Growth responses of the *arg⁻* to intermediates of the arginine pathway

Mutants	Medium supplement			
	None	**Ornithine**	**Citrulline**	**Arginine**
a, d, h, j, *o, p, y, z*	no growth	growth	growth	growth
c, e, f, g, *i, k, m, n,* *q, r, s, t,* *u, v, w, x*	no growth	no growth	growth	growth
b, l	no growth	no growth	no growth	growth

2. Accumulation of intermediates

Another test, the ability of mutant strains to accumulate intermediates of the pathway, is more discriminating. A mutant will not make the product of the enzyme it lacks. However, it is likely that the mutant will accumulate the intermediate just *before* the genetic block and possibly others before that. If we look at the two mutants (*b* and *l*) that grow only on arginine, but not on citrulline or ornithine, we may ask whether either accumulates the intermediate argininosuccinate (Fig. 17–1). Indeed the *arg-10* mutation (*l*) accumulates both argininosuccinate and citrulline, while the *arg-1* mutation (*b*) accumulates only citrulline. This finding is consistent not only with the nutritional behavior of the mutants, but also with the observation above that they complement with one another and assort independently in meiosis. Plainly, the two mutations lie in different genes, encoding different enzymes.

Accumulation of intermediates of a metabolic pathway can be put to clever use in **cross-feeding** experiments. The intermediate accumulated by one mutant strain may be used by another mutant strain as a nutritional supplement, assuming the intermediate can penetrate the cell membrane. In this way, geneticists can often order mutations in a pathway from the earliest to the latest by growth tests. In a pathway with unknown intermediates, the purification and characterization of the growth-supporting chemical accumulated by one strain and used by another will lead to the identification of these chemicals and ultimately of the enzymes that constitute the pathway. Many metabolic pathways were elucidated in this way, based on work with microorganisms.

3. Enzymological characterization

Assay of the arginine biosynthetic enzymes in the mutants reveal several new points. First, not all enzymes are represented by mutants. This is not surprising; in this group of 26 mutants, it is not highly probable that all of ten different enzymes would be "hit" by mutation. Indeed, some mutants are the sole representative of their gene. Second, somewhat surprisingly, two complementation groups (III and V in our example, Fig. 17–5) affect the same enzyme (No. 9). Further investigation shows that this enzyme, carbamoyl phosphate synthetase, is made up of two distinct polypeptides, and that a corresponding gene (*arg-2* or *arg-3*) encodes each one. The two genes, in fact, assort independently. Many examples of this relationship are known, and for that reason the "one-gene-one-enzyme" slogan was superseded by the one gene-one polypeptide formulation that holds to this day.

4. Epistasis

We have already encountered the term epistasis in Chapters 6 and 7. Formally, epistasis denotes the ability of an allele of one gene to mask the allelic state of another when they coexist in the same genotype (e.g., in a double mutant). We have seen some biochemical scenarios of epistatic relationships in Chapter 7; a more blunt example might be a mutation to baldness that is necessarily epistatic to a gene (any gene) that determines hair color.

By the same token, a gene early in a biochemical pathway will be epistatic to the genes later in a pathway with respect to the intermediates accumulated. However, a later mutation (i.e., farther down the pathway) will be epistatic to an earlier one with respect to the intermediates that a double mutant will grow on. Therefore, one must be cautious in generalizing about what epistatic relationships tell us about the order of function; it depends upon which phenotype is analyzed. By clever use of epistatic relationships, the order of function of genes affecting extremely obscure steps in cellular processes can be inferred.

F. Analysis of Complex Processes

The methodology outlined above for the arginine pathway is applicable, in any organism, to complex functions such as the cell division cycle, the steps of macromolecular synthesis, the secretion of proteins from the cell, the course of meiosis, or the course of developmental repertories. In all cases, the methodology begins with selection of a large number of mutants that share a phenotype indicating a defect in a particular process. They are grouped by complementation tests, and each group is characterized by what stage of the process it might affect, and what specific function in that stage has been blocked. By this means, a genetic dissection of many complex processes can be accomplished, with the ultimate identification and characterization of most of the proteins required.

QUESTIONS AND PROBLEMS

1. Why is a test for complementation in *Neurospora* not confused by wild-type recombinants arising from the mutants tested? Answer the same question with respect to yeast and *Drosophila*.

2. Three mutant strains of *Neurospora* (X, Y, and Z) that required histidine are given to you. All have the same mating type. A test for complementation shows that X and Y complement with one another, but Z fails to complement with either. Because you do not know whether strain Z is a single mutant or a double mutant, you cannot interpret the results. Frame the one- and two-gene hypotheses clearly, and show how you would distinguish them with crosses.

3. What is the difference between complementation and recombination tests for allelism, and what does each one tell us?

4. A histidine-requiring strain of *Neurospora* was mated to wild type, and the progeny were 75% histidine-requiring and 25% wild type. This curious ratio could not be explained by poor germination of the wild type spores, because germination was 95%. Give a simple explanation of the result.

5. A set of 7 *Neurospora* mutants, (*a–g*), was tested for complementation in heterokaryons. It gave a rather complex pattern, as follows:

    a

    a 0 b

    b + 0 c

    c + 0 0 d

    d + 0 + 0 e

    e + + + + 0 f

    f + 0 + + + 0 g

    g + 0 + 0 + 0 0

    How many genes appear to be represented by the mutants? What mutants represent each gene? What is the basis for the complex pattern of complementation?

6. A newly discovered sugar, zotobiose, was found as a minor component of cell walls of *Neurospora*. Zotobiose-requiring mutants were isolated, and this told us that zotobiose is essential for growth. The questions that follow apply to a set of 12 *zot⁻* mutants.

    (a) Pairwise complementation tests yield the data below in a complementation grid (+ = growth, O = no growth of the heterokaryon in a zotobiose-free medium). Give the number of genes involved, and identify the alleles of each gene. Identify intragenic complementation if it occurs.

```
  a
a 0 b
b + 0 c
c 0 + 0 d
d + + + 0 e
e + + + + 0 f
f 0 + 0 + + 0 g
g 0 + 0 + + 0 0 h
h + + + 0 + + + 0 i
i + + + + 0 + + + 0 j
j + + + + 0 + + + + 0 k
k 0 + 0 + + 0 0 + + + 0 l
l 0 + 0 + + 0 0 + + + 0 0
```

(b) Certain *zot* mutants accumulate related compounds when they are starved for zoto-biose. Investigators hoped that such compounds might be precursors of zotobiose. Three compounds, A, B, and C were in fact found, and their appearance in single and double mutants was as follows:

Single mutants	Compound(s) accumulated	Double mutants	Compound(s) accumulated
f	none	f, h	none
h	B	e, h	B
e	B, C	b, h	B
b	A, B, C	e, b	B, C
		f, e	none
		f, b	none

Place the mutants *f, h, e,* and *b* on a metabolic map of zotobiose synthesis that shows the order of the compounds A, B, C, and zotobiose.

(c) The mutants *b, f, h, i,* and *j* were crossed in various pairwise combinations, and the number of wild-type (*zot*+) recombinants was determined for each cross. The results were as follows:

Cross	Wild-type recombinants
$i \times j$	2×10^{-5}
$f \times h$	25%
$f \times i$	3%
$h \times j$	25%
$f \times b$	7%
$b \times j$	4%
$b \times h$	25%

Interpret these data in the form of a genetic map, indicating which mutations independently assort, and the linkage of any linked mutations with gene order and map distance.

(d) In the cross of $f \times i$ in question 6(c), above, assuming no double crossovers, what percentage of tetrads would be tetratypes? Give your reasoning.

(e) In mapping mutant h with the metabolically unrelated, but linked mutations r and t, the following cross was done: $h\ R\ t \times H\ r\ T$. The progeny genotypes and frequencies are shown below. Give a genetic map of the three genes.

Progeny genotypes	Number
H R T	25
h R T	20
H r T	724
h R t	701
H r t	10
h r t	20

7. How would you embark upon the analysis of the yeast cell wall with genetic techniques?

Chapter 18

BACTERIAL AND BACTERIOPHAGE GENETICS

Prokaryotes exchange genes in several different ways, all quite different from the mating mechanisms of eukaryotes. Moreover, recombination mechanisms are much less regular. For these reasons, geneticists have developed specific methods for bacterial and bacterial viruses. From these studies, a great deal has been learned of the basic biology, as well as the genetics of prokaryotes. Because of the large numbers of progeny in prokaryotic crosses and the selectivity possible in finding recombinants, the most detailed genetic studies of individual genes have been done in these organisms.

A. Mutations in Prokaryotes

The genetics of prokaryotes is based largely on the use of metabolic mutants. Three types of such mutants are (i) auxotrophs, which require a metabolite not needed by the prototrophic wild type; (ii) antibiotic-resistant mutants, which can tolerate common antibiotics like penicillin and streptomycin; and (iii) utilization mutants, which differ from the wild type by their ability or inability to use specific compounds as carbon or nitrogen sources for growth. A more general class of mutants are conditional mutants. The most common type of conditional mutation is temperature-sensitive. In *E. coli,* mutants that can grow at 30° C, but not at 37° C (the normal laboratory temperature for this bacterium), carry mutations that affect the heat stability of the protein encoded by the affected gene. Because temperature-sensitive mutations can occur in any gene, geneticists have been able to mutate genes with "indispensable functions." Mutations that wholly blocked the function of such proteins, such as DNA polymerase or a ribosomal protein, or enzymes of cell wall assembly, would not be able to survive under any condition. Conditional mutations give us access to the study of these genes. With all types of mutant, the isolation, gene-transfer and recombination experiments are performed by using selective media, or other conditional regimes, which discriminate mutant and wild type cells. Bacterial genetics has developed most fully with the use of the common enteric bacterium, *Escherichia coli,* its relatives, and its viruses.

Geneticists have isolated many mutants by negative selection with a method called **penicillin enrichment.** The antibiotic, penicillin, blocks formation of the resilient cell wall of many kinds of bacteria. However, the cells themselves do not stop growing in penicillin; they continue to do so until the wall becomes too fragile to contain the internal pressure (turgor) of the cells. After a period of growth in the presence of penicillin, therefore, growing cells burst and die. Good use has been made of this phenomenon in selecting cells that will not grow, such as auxotrophs in the absence of their required nutrient. Consider a large, prototrophic bacterial population that is irradiated to induce mutations and is then allowed to grow in minimal medium containing penicillin. The population will become enriched in mutant cells, which survive with higher probability, much like the enrichment of mutant *Neurospora* cells during growth and filtration described in Chapter 17. By washing the bacterial cells free of penicillin and spreading them on plates of medium containing a nutrient needed by a particular class of auxotrophs, an investigator can obtain colonies, a substantial number of which are of the desired type. The isolation of mutants is further facilitated by the replica plating method. As noted in Chapter 16, replica plating involves printing colonies from one plate onto another without disturbing their arrangement. The method allows rapid identification of auxotrophs after penicillin enrichment, and the counting of different progeny in matings.

B. Bacterial Gene-transfer Mechanisms

Table 18–1 lists and defines the major bacterial gene transfer systems.

1. Transformation

Bacterial transformation, described in Chapter 1, includes (i) binding and uptake of **donor** DNA by **competent,** recipient cells; (ii) integration of one or more segments of this DNA into the chromosome of the recipient; and (iii) loss, by degradation, of excess DNA (Fig. 18–1). In practice, geneticists expose a population of recipient cells having a mutation to the DNA of a strain having the wild-type allele. The donor DNA is fragmented upon isolation into pieces usually no more than 1% the length of the chromosome of the donor cells. After a time for DNA uptake, the recipient population is spread on a medium that selects transformed cells, and against the untransformed phenotype. Automatically, transformants that have integrated the wild-type allele from the donor DNA will grow on the new medium. The same technique is used to select for antibiotic-resistant transformants in an antibiotic-sensitive, recipient population. The use of double mutants as recipients reveals in some cases a high rate of **cotransformation** of both genes. This shows that the two genes lie near to one another on the bacterial chromosome, and are therefore borne on the same small fragment of DNA. Cotransformation is indicated if the two markers simultaneously transform the same competent cell at a frequency greater than the product of their individual transformation frequencies among competent cells. This is the most direct display of molecular linkage in bacterial recombination.

Table 18–1   Bacterial recombination processes

Process	Variant	Description
Transformation	Linear DNA	Entry of naked DNA into cell, followed by integration into chromosome.
	Plasmid	Entry of naked circular plasmid into cell, followed by replication of plasmid.
Conjugation	F⁺ × F⁻	Only the F plasmid enters F⁻, leaving a copy in F⁺.
	Hfr × F⁻	F factor in chromosome directs attached bacterial DNA into F⁻ cell by replicative transfer.
	F′ × F⁻	F plasmid containing bacterial genes enters F⁻ cell as in F⁺ × F⁻ mating.
Transduction	Generalized	Bacterial DNA in phage protein coat is injected into recipient, followed by integration into chromosome.
	Specialized	Bacterial DNA attached to phage DNA enters during infection by phage, integrates into chromosome at phage attachment site.
Transposition	General	Insertion sequence or transposable element transfers itself within cell to another location on DNA of cell. May be replicative.

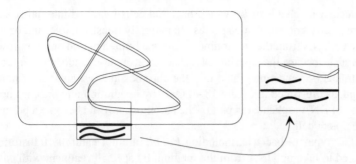

Binding and entry

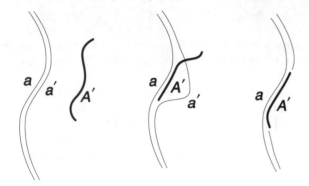

**Integration of donor DNA as heteroduplex
with recipient DNA**

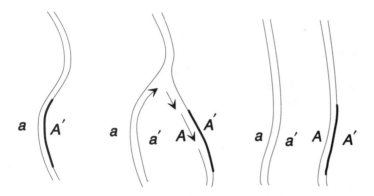

**Replication of heteroduplex recipient chromosome
forming homoduplex transformant (selection can
now be applied for A gene function).**

Figure 18–1 Steps in bacterial transformation.

The molecular mechanisms by which bacterial cells bind and internalize large DNA molecules are poorly known. Some species are naturally transformable, and become competent to take up DNA from the environment during a limited period of their growth in culture. In most such species, the entering DNA becomes single-stranded as it enters, and one of the two strands becomes integrated into the chromosome by replacing a homologous, resident strand, as illustrated in Figure 18–1. Only by integration into the chromosome can a simple piece of DNA replicate (as part of the chromosome) and become a permanent feature of the recipient cell.

Many other types of bacteria, including *E. coli,* are not naturally transformable, but may be induced to take up DNA from the medium by special treatment with calcium and heat. Transformation by this means begins with the uptake of double-stranded DNA, and the procedure allows a geneticist to introduce entire, double-stranded, circular plasmids into cells. Because plasmids have a replication origin that is recognized by the bacterial replication system, they may replicate without integrating into the chromosome.

2. Conjugation

The bacterium *E. coli* has a well-developed sexual system by which DNA is transferred from one cell to another, followed by the integration of this DNA into the chromosome of the recipient.

a. The F plasmid and its transfer.   The bacterial sexual system depends upon a special, large plasmid called the F (for fertility) factor, about 2% the size of the *E. coli* chromosome. Strains with an autonomous F factor are called F$^+$, and those without it are called F$^-$. The F factor, which is usually separate from the bacterial chromosome, encodes proteins required for the transfer of the F factor to another cell. Among these proteins are those of the **pili** (sing., pilus), specialized, hair-like appendages on the cell exterior. Pili can establish contact and adherence to F$^-$ cells, and the F$^+$/F$^-$ pair of cells then forms a **conjugation bridge** between them. The F plasmid then directs its transfer to the F$^-$ cell (Fig. 18–2, left). As the F plasmid replicates in the rolling circle mode, one copy enters the recipient cell through the conjugation bridge, and the other is retained by the donor cell. When an entire plasmid length of DNA has entered, it is rendered circular and can thereafter replicate once each time the recipient cell divides. The F factor that enters may mobilize its further transfer to still other F$^-$ cells. A mixture of F$^+$ and F$^-$ cells will therefore quickly become uniformly F$^+$, a process called **infective conversion.** In this process, no genes of the bacterial chromosome are transferred.

Many plasmids able to transfer to other cells exist among bacteria, and some of them carry antibiotic-resistance genes. Some plasmids can even transfer between related species of bacteria. These plasmids are a troublesome feature of hospital settings, because they can distribute antibiotic resistance widely among bacteria found there. The F factor, however, does not carry antibiotic-resistance genes.

b. Chromosome transfer.   Some cells can transfer bacterial genes to F$^-$ cells. In these cells, the F factor, instead of being a separate, circular plasmid, has integrated into the circular chromosome of the host. Integration takes place by a recombinational step, effectively a single crossover between two circular DNA molecules (Fig. 18–3). The F factor may integrate at any of several positions on the bacterial chromosome, and in either orientation. Cells with an integrated F factor are called Hfr (*high frequency of recombination*) cells. The F factor replicates as a part of the bacterial chromosome. However, the integrated F factor can now transfer not only its own DNA, but also the bacterial chromosome to which it is attached (Fig. 18–2, right). When an Hfr cell begins to transfer DNA to an F$^-$ cell, a part of the F factor enters first, followed by the bacterial chromosome adjacent to it. We call the initial bacterial DNA the **origin of transfer.** The process of transfer is governed by replication of the large, circular bacterial chromosome in the rolling-circle mode, initiated and directed by F DNA. The amount of bacterial DNA transferred is determined by the time at which, through random movements of the conjugants, the delicate conjugation bridge breaks.

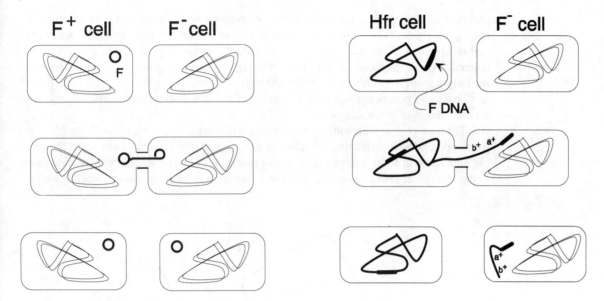

F⁺X F⁻ mating: transfer of F plasmid
(infective conversion)

Hfr X F⁻ mating: transfer of bacterial chromosome, mobilized by integrated F factor

Figure 18–2   Conjugation of F⁺ × F⁻ cells (left) and of Hfr × F⁻ cells (right).

In Hfr × F⁻ matings, transfer of an entire Hfr chromosome into a recipient cell takes 100 minutes at 37° C. By that time only a few mating cells of a population remain together. Therefore, one recovers many recombinants, called **exconjugants,** carrying genes close to the origin of transfer, and very few carrying more distant genes. Almost none (less than 0.1%) of the exconjugants have the Hfr phenotype, because a complete copy of the F factor enters F⁻ cells only after the whole of the chromosome of the Hfr donor has been transferred. In conjugation experiments, a population of exconjugants is spread on media that counterselects both types of parental cells. The medium is designed to support the growth only of the F⁻ cells with particular genes transferred by the Hfr. If, among exconjugants carrying one Hfr marker, there are more than 50% carrying another Hfr marker, linkage of the two markers is indicated. However, recombination is so efficient in *E. coli* that genes only a short distance apart physically will show little or no genetic linkage: the connection they might have on the Hfr chromosome is broken up and randomized during the integration process. For that reason, another mapping method has been devised, and will be described below.

Donor Hfr cells remain genetically unchanged after mating, because they transfer chromosomes by replicating them and keeping a copy. The recipient F⁻ cells almost always remain F⁻ because only part of the F factor is transferred initially, and even this fragment is lost during the integration process in the recipients.

Notice how different this mating mechanism is from that of eukaryotes. Instead of fusion of gametic cells and nuclei, bacterial matings consist of a polarized transfer of DNA from a donor to a recipient, both of which retain their integrity.

c. The time-of-entry mapping method.   Agitation of mating cultures in a blender disrupts contact between conjugating cells. This offers an alternative means of mapping genes, called the **time-of-entry** method. The process starts at "time zero," when Hfr and F⁻ cells are mixed. The Hfr cells generally have wild-type alleles of auxotrophic mutations, and are mixed with F⁻ cells that carry the auxotrophic alleles. This makes it easy to select for prototrophs (Hfr markers) among exconjugants, and to eliminate the original F⁻ cells. In addition, parental Hfr cells are chosen in practice to be sensitive to an antibiotic to which the F⁻ cells are resistant. This allows selection against the original Hfr parent. Samples of the mating mixture are withdrawn periodically and agitated to break conjugating pairs. The samples are then plated on a medium containing the antibiotic and lacking one of the nutrients required by the F⁻ strain. The choice of the latter may depend upon the purposes of the experiment; in any case, the plating is highly selective for only certain recombinants. Unlike analysis of progeny of eukaryotic crosses, therefore, bacterial geneticists see only a fraction of the potential recombinant progeny, and no true "parental" cells in crosses.

The time-of-entry mapping method measures the time at which a given Hfr marker appears among the exconjugants. For any gene of interest, one removes samples of the mating population at short intervals (e.g., every two minutes), separates the conjugating pairs in the blender, and, after plating the mixture, determines the increasing number of excon-

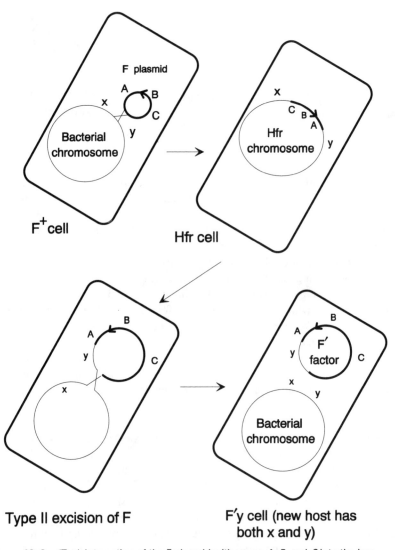

Figure 18–3   (Top) Integration of the F plasmid with genes *A, B,* and *C* into the bacterial chromosome between bacterial genes *x* and *y.* Integration requires a single recombination event; note the resulting gene order. The origin of transfer is indicated by the arrow head; the order of transfer during conjugation of the Hfr with an F⁻ cell is B-C-x. (Bottom) Imperfect excision of the F DNA from the Hfr chromosome, incorporating bacterial gene *y* into the F′ DNA. At the lower right, the F′ DNA is shown after transfer to a cell that has all genes, including *y;* the resulting cell is a merodiploid for gene *y.*

jugants that have acquired the Hfr marker in question. Soon after a marker begins to appear among exconjugants, the number of cells with that marker reaches a plateau. At that time, all pairs that were mating at the time of blending have finished the transfer of that marker. A marker close to the origin of transfer will begin to appear in exconjugants very soon after the parents are mixed; those far from the origin of transfer must wait until all earlier markers have entered the F⁻ cells. For any given marker, the time of entry is determined by extrapolating the rising numbers of selected cells back to the time axis (Fig. 18–4); that is, to the time when the marker first begins to enter F⁻ cells. The initial times of entry then become a map of the genes on the chromosome—**a map measured in minutes, not recombination percentages.**

The chromosomal position of the F factor is the same in all cells of a given Hfr strain, but different for different Hfrs. Comparison of different Hfrs for the markers that they donate first in matings reveals (i) that the F factor may have different chromosomal positions; (ii) that the F factor may be integrated in either orientation (the bacterial chromosome can be transferred in either polarity); and (iii) that the bacterial chromosome is circular: genes at two ends of the sequence transferred by one Hfr strain are closely linked in another.

d. F′ plasmids. The integration of the F plasmid into the chromosome that yields an Hfr cell (Fig. 18–3) is a rare event. The F factor may also occasionally **excise** from the chromosome of an Hfr cell. In some cases, this may exactly reverse the original integration event, but it is more often an imperfect process. In the latter case, the excision process removes not only the F DNA but also bacterial genes on one or the other side of it. The excised plasmid that emerges is circular, and is called an F′ factor. This factor is transferred from F′ cells to F⁻ cells as easily as the normal F factor. When an F′ factor enters a normal F⁻ cell, the resulting F′ cell is diploid for the bacterial genes carried on the F′ factor (Fig. 18–3). These partial diploids, called **merodiploids**, offer geneticists the opportunity to study dominance relationships in bacteria, which are normally haploid.

3. Transduction

In transduction, the protein coats of bacteriophages (phages) mediate transfer of DNA from a donor bacterium to a recipient bacterium. The phenomenon depends upon a class of phage called **temperate phages**. These phages, while they can infect, multiply, and destroy their host cells (*see* Chapter 2), can also have a latent existence in their hosts. This state, represented only by the DNA of the phage and called a **prophage**, may be a circular

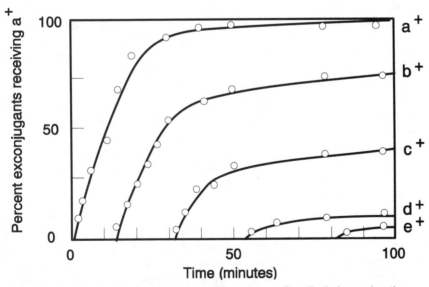

Figure 18–4 Kinetics of transfer of different Hfr markers to F⁻ cells during conjugation. The ordinate is the fraction of the exconjugants that have also received and integrated the early marker, *a*⁺.

plasmid (e.g., phage P1 of *E. coli*) or, more commonly, a segment of the bacterial chromosome at a particular site (e.g., phage λ [lambda] of *E. coli* or phage P22 of *Salmonella typhimurium*). The latter form of prophage DNA arises by the integration of circular phage DNA into the bacterial chromosome, similar to F factor integration in the formation of an Hfr cell.

UV irradiation or treatment with heat or certain chemicals that block DNA replication may induce prophages to multiply and **lyse** (destroy) their hosts. Soon after the onset of the "lytic cycle," as this process is called, the bacterial chromosome breaks down, and occasional fragments later become encapsulated in phage protein coats. When bacteria lyse, liberating progeny, some phage coats carry bacterial DNA. These are rare **transducing phages,** vehicles that inject their segment of bacterial DNA into new recipient hosts just as they would have injected phage DNA if they were normal phage particles. The recipients of particular bacterial genes (transductants) can be selected on media on which the untransduced cells cannot grow. For instance, prophage induced to multiply in a bacterium carrying the his^+ allele of a gene required for synthesis of the amino acid histidine can transduce a his^- host. Only cells that have acquired the his^+ DNA will give rise to colonies when the newly infected recipient population is plated on a medium lacking histidine.

Bacterial recipients that carry a prophage are called **lysogenic** cells. Lysogenic cells do not permit the replication of superinfecting phages of the type they carry as a prophage, even if phage DNA enters the cell. Such recipient bacteria are used in transduction experiments so that normal phage particles—present in large numbers—do not infect and kill potential transductants. However, non-lysogenic cells may be employed if the phage used is a mutant which cannot grow under the conditions of infection in the transduction experiment.

Once inside the cell, the fragment of "donor DNA," the DNA injected by a transducing phage particle, will recombine with the recipient chromosome with high probability. It does so by mechanisms that integrate transforming DNA or the DNA transferred by conjugation. The limiting factor in transduction is the small number of transducing phages (carrying DNA from the previous bacterial host) in the large population of normal phages. As a rule, only about 1/1000 of the particles in a transducing phage population carry bacterial DNA, and only 1/50 of those carry any particular gene. Because very large populations of bacteriophage (over one billion) are available in normal practice, however, this is not a severe limitation.

Bacterial transduction has two forms. In **generalized transduction,** using the phages P1 and P22, for instance, the rare transducing phage particles carry only bacterial DNA and no phage DNA (Fig. 18–5). In these phages, phage coats encapsulate only one molecule of DNA of a specific length—either a fragment of bacterial DNA or one phage DNA molecule. In **specialized transduction,** the bacterial DNA is attached to a large portion of phage DNA (Fig. 18–5). The kinds of phages that undergo specialized transduction are those like the phage lambda (λ) in which the encapsulation process must begin with the recognition of phage DNA. The reason that bacterial DNA is occasionally included is because it is attached to the mature phage DNA. These joint molecules arise through occasional accidents in prophage excision from the host chromosome upon induction, similar to the formation of F′ plasmids. Because specialized transducing phages can include only bacterial genes that are adjacent to the integrated prophage, only those genes can be transduced; hence the term specialized transduction.

Bacterial transduction is most important in fine-scale mapping of bacterial mutations. The time-of-entry method of conjugational mapping permits mapping of genes no less than about two minutes apart (2% of the 100-minute bacterial chromosome). Genes closer together must be mapped by a recombinational method. The fragments carried by generalized transducing phage particles are, conveniently, around 2% the length of the bacterial chromosome. Therefore, two markers that are **cotransduced** must be physically close to one another on the bacterial chromosome. With enough markers in a short region, the region can be mapped with high precision, cotransduction being less frequent for those markers farther

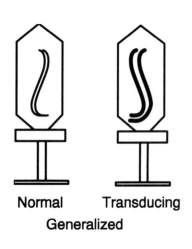

Normal   Transducing
Generalized

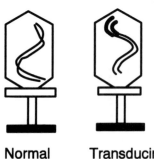

Normal   Transducing
Specialized

Figure 18–5 Generalized and specialized transducing phages. In each case a normal and a transducing phage are shown. Heavy lines indicate bacterial DNA.

apart, and more frequent for those closer together. (We will return to this methodology below.) The molecular phenomena underlying the recombinational events of the integration process involve both breakage and rejoining of the homologous DNAs and the associated gene conversion and repair described in Chapter 10 for eukaryotic recombination.

4. Transposition

Transposition (Fig. 18–6) refers to the transfer of "mobile" DNA *within* a cell from one position to another. Transposition of the mobile DNA can be confined to a single DNA, or it can take place between different DNA elements, such as the bacterial chromosome, plasmids, or phage DNAs. The mobile DNA elements are called **transposable elements.** Every transposable element is physically a segment of a larger DNA. The ends of a transposable element have small, specialized DNA sequences that point in opposite

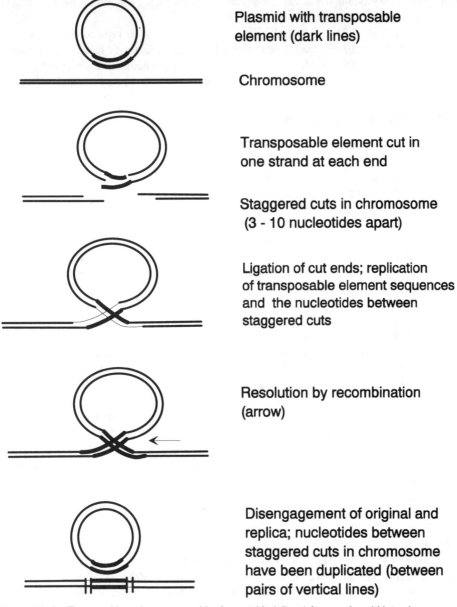

Plasmid with transposable element (dark lines)

Chromosome

Transposable element cut in one strand at each end

Staggered cuts in chromosome (3 - 10 nucleotides apart)

Ligation of cut ends; replication of transposable element sequences and the nucleotides between staggered cuts

Resolution by recombination (arrow)

Disengagement of original and replica; nucleotides between staggered cuts in chromosome have been duplicated (between pairs of vertical lines)

Figure 18–6   Transposition of a transposable element (dark lines) from a plasmid into chromosomal DNA.

directions; that is, they are **inverted repeats.** Transposition occurs through the action of an enzyme called a **transposase,** encoded by a gene in the transposable element. The transposase makes use of the inverted repeats to transfer them *and* all the DNA between them (that is, the whole transposable element) to another position. Many transposable elements replicate as they do so, leaving a copy in the original position.

During transposition, the recipient DNA suffers **staggered cuts** from three to ten nucleotides apart on the two strands. Ultimately, the shorter chain is extended at each end by DNA polymerase to equal the length of the longer one as the transposon inserts itself between the cut ends (Fig. 18–6). This results in a characteristic duplication of the recipient DNA, one copy just beyond each end of the transposable element.

The simplest transposable elements are called **insertion sequences,** having only a transposase gene in their DNA. (Many have defective transposase genes and rely on other elements in the same cell for the enzyme.) More complex **transposons** have insertion sequences, inversely or directly repeated and often highly modified, at each of their ends. Transposons usually, but not always, move in their entirety, but sometimes the insertion sequence at one end moves independently.

The term "selfish DNA" describes these elements best: their replication is under their own control; they generally do no good to the host, and in fact may harm it by disrupting genes; and they may multiply faster than cells, increasing the number of copies per cell over time. Higher organisms have large numbers of these repeated sequences in their DNA.

C. Mapping Bacterial Genes

Mapping a gene identified by a new mutation has two phases. The first is localizing it roughly on the bacterial chromosome by conjugation experiments. The second is transductional mapping, in which it is mapped as precisely as nearby markers will permit. The importance of the latter phase becomes clear when we recognize a general difference between prokaryotic and eukaryotic gene organization. Genes of related function, such as those encoding enzymes of a biochemical pathway, are scattered about the genome in eukaryotes, whereas such sets of genes in bacteria frequently lie together at a particular position on the bacterial chromosome (*see* Chapter 19). The map order has some importance to studies of gene expression, and consequently, transductional analysis plays a major part in initial descriptions of such gene arrays.

1. Conjugational mapping

Let us consider a group of genes responsible for the synthesis of a fictitious, essential compound we will call malamine. In a mutant hunt, mutants requiring malamine (*mam*⁻) were isolated, starting with F⁻ cells, by the penicillin enrichment method. In the first step of rough mapping, we could use either of two related methods. In one, different *mam*⁺ Hfr strains are mated to the *mam*⁻ mutant. The Hfr strains are chosen to represent early transfer of *different* regions of the bacterial chromosome. We look for the Hfr strain that yields large numbers of *mam*⁺ exconjugants. This identifies the segment of the bacterial chromosome, transferred "early" by the Hfr strain (the first 20 minutes of mating), carrying the *mam*⁺ marker. While other Hfrs might yield Mam⁺ exconjugants, they would be appreciably fewer, owing to the orientation of the F factor or the distance between the origin of transfer and the *mam*⁺ gene in question.

A "panel" of F′ strains provides an alternative method of roughly localizing the *mam* gene. Various F′ strains have been isolated in which the F DNA carries different segments of the bacterial chromosome. Each of these segments is transferred efficiently, by infective conversion, to F⁻ cells. The F′ strain that transfers a *mam*⁺ marker to the *mam*⁻ cell identifies the segment of the bacterial chromosome on which the *mam* gene lies.

The next step in roughly mapping the *mam* gene is an interrupted mating experiment with the appropriate Hfr strain. Here, it is desirable to use an F⁻ recipient with the streptomycin-resistance marker (which can be transduced into the strain), and the Hfr and F⁻

strains should, if possible, differ with respect to other markers in the region known to harbor the mam^- mutation. (The streptomycin-resistance gene cannot be in the same region as the mam gene. Suitable counterselection markers are available in other regions of the chromosome if streptomycin resistance cannot be used.) The time of entry is determined in a mating of Hfr str^S $mam^+ \times F^-$ str^R mam^-, selecting for cells that are str^R mam^+. If other, nearby markers are used, their times of entry can be compared in the same experiment to that of mam^+. This procedure can localize the mam gene in question to a 2-minute segment of the bacterial chromosome.

2. Cotransductional mapping

We now use transduction to map the mam gene more precisely. The *E. coli* chromosome is well marked, and many mutations are available for most regions of the chromosome. Therefore, we wish to use a mam^+ P1 lysogen carrying wild-type alleles of genes (which we will call, in order, a^+, b^+, and c^+) in the region of the mam gene and use it as the donor, and the mam^- mutant (having mutant alleles a^-, b^-, and c^-) as the recipient. By selecting for mam^+ transductants on a medium that will support the growth of the a^-, b^-, and c^- mutants, we determine, with replica plating, how many also received a nearby wild-type a, b, or c marker from the donor. These transductants would be cotransductants for mam^+ and one or more of the other markers, a^+, b^+, and c^+. Consider the following cotransduction percentages for markers a, b, and c: *a - mam:* 2%; *b - mam:* 60%; and *c - mam:* 20%. Plainly, given the order a-b-c, mam is closer to b than to the other two markers, but we do not know which side of b it is on unless we also compare the a - b and the b - c cotransduction frequencies. In separate experiments, these are *a - b:* 10% and *b - c:* 30%. This allows us to give the order as a - b - mam - c, because mam is more distant than b from a, and mam is closer than b to c.

3. Three-point transduction

Let us now suppose a number of different mam mutants all lay in the same general position, between b and c. Because they were isolated independently, it is unlikely that they represent the same mutational site, or even the same gene, knowing as we do that bacterial genes of related function often form clusters. The identity or non-identity of mam^- mutants can be established by performing transductions between different mam mutants: if mam^+ transductants appear in a $mam^- \times mam^-$ transduction, the mutations cannot lie at the same site on the bacterial chromosome, although they may lie in the same gene. (We will not test the mam mutants for allelism here; techniques are readily available for such a test.) Our task is now to try to order the mam mutations (for example, $mamC$ and $mamD$) with respect to one another and with respect to an "outside" marker, such as b.

Mapping now requires reciprocal three-point crosses, in which one takes advantage of the greatly differing probabilities of double and quadruple crossovers. In transduction, the integration of a donor marker requires an even number of crossovers, at least one on either side of the marker. Consider the markers b, $mamC$, and $mamD$ with the two possible orders b-$mamC$-$mamD$ and b-$mamD$-$mamC$. A pair of crosses is performed, both having a b^+ donor and a b^- recipient. The members of the pair differ in that the donor/recipient relation is C^+D^-/C^-D^+ in one, and C^-D^+/C^+D^- in the other. The frequency of b^+ among Mam$^+$ ($mamC^+mamD^+$) transductants is determined. Let us say, in this example, that the outcomes of the crosses were:

    1. Donor b^+ $C^+D^- \times$ Recipient b^- C^-D^+: 20% b^+ among Mam$^+$

    2. Donor b^+ $C^-D^+ \times$ Recipient b^- C^+D^-: 1% b^+ among Mam$^+$

Because integrating b^+ is less frequent in the second configuration than in the first, we conclude that in the first cross, integration of b^+ in the mam^+ class requires only a double crossover, while in the second, a quadruple crossover is required. One must then put the markers in the order (*b-mamC-mamD*) that is consistent with this inference:

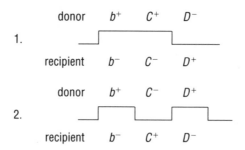

A trained eye would see that b^+ and C^+, when together in the donor DNA, cotransduce much more frequently than when b^+ and D^+ are together in the donor DNA. Thus C is nearer to b than D is.

Try the order *b - mamD - mamC* to confirm its poor fit to the data. Note that b cannot be medial because of the initial conditions: b is an "outside" or flanking marker.

D. Bacteriophage Genetics

1. General genetic rationales

Phages may have DNA or RNA as their genetic material, wrapped in a protein coat. The mature particle is known as a **virion,** and as such it is quite inert. Small phages may be very small indeed, containing as few as 3 genes, as in the RNA viruses, or 10 to 11 genes, as in small DNA viruses. Other phages may have 40–50 genes, as in phage λ or as many as 150 genes, as in phage T4. Phage DNA may be linear or circular, and either single- or double-stranded, although the single-stranded phages always have a double-stranded stage during their replication. Phages have particular properties of virulence, the range of hosts that they can infect, their rate of replication, and the damage that they do to their hosts. They may be **virulent,** without any latent (prophage) state, or **temperate,** having an optional prophage state in their hosts (Table 18–2). We have described some of the characteristics of temperate phages in connection with bacterial transduction.

An infection by a typical large, virulent phage such as phage T4 of *E. coli* starts with the injection of phage DNA into the cell, followed soon by the replication of this DNA. After injection of the DNA, the protein coat of the virion is no longer necessary for the infection to proceed. (This was the basis of an early experiment that demonstrated that DNA could carry all the information necessary to produce more phage, including the protein

Table 18–2 Virulent and temperate bacteriophage

Phage type	Behavior
Virulent	Upon injection, replication of phage DNA ensues, followed by encapsulation in phage coats. The host lyses and liberates as many as to 250 phage particles per cell.
Temperate	Lytic cycle: Upon injection, phage multiplication and lysis occurs as in virulent phage.
	Lysogeny: After injection, phage DNA, adopting a circular form, integrates into the bacterial chromosome at a phage attachment site and becomes latent, replicating as a part of the chromosome. Some temperate prophages maintain themselves as separate plasmids. Both integrated and plasmid forms are called prophages, kept dormant by a phage-encoded repressor protein. Prophages may be activated to undergo the lytic cycle with ultraviolet light, heat, or drugs that interfere with DNA replication or integrity.

coat.) Once in the bacterial host cell, phage DNA is transcribed and translated by the molecular machinery of the host. However, the DNA of this large phage encodes many proteins required in the infection, such as those that inactivate host functions and break down the large bacterial chromosome. It also encodes certain metabolic enzymes needed to make special, modified nitrogenous bases used by the phage. In this way, the phage subverts the cell, directing much of the cell machinery to the production of bacteriophage. The bacteriophage DNA also encodes a protein that attacks the bacterial cell wall, leading to lysis and the escape of progeny phage at the end of the infection. Finally, the phage DNA encodes the proteins of the virion, which encapsulate the DNA in the latter half of the phage infection.

Methods of phage genetics rely on the study of the numbers and types of **plaques** that appear on bacterial lawns. Plaques are initiated by single phage particles. If a dilute suspension of phages, mixed with a large number of bacteria, is spread on a fresh medium, the uninfected cells form the lawn, and the few infected cells are killed. The progeny phages of this initial infection will infect nearby cells in the lawn and kill them, too, an infection that expands to form a zone of clearing, or a plaque. Phage particles in a suspension are counted in this way. However, the plaques of mutant phages may have phenotypes differing in size, turbidity and outline, and phages of different genotypes may differ in the bacterial strains they can infect. The last are called host-range mutants.

Recombination of mutant phages can be studied and measured in mixed infections. Here, phage differing in two or more characters (host range and plaque size, for instance) are added to a bacterial population. Each phage is added in numbers (e.g., 5 per bacterium) that assure that every cell will be infected by at least one of each type. Upon lysis of the mixed-infected population, progeny phage are plated in conditions that reveal the allelic differences between the parent phages. The percentages of recombinant phage are the basis of linkage maps, made by procedures with which you are already familiar. The large number of phage particles and the high rates of recombination in phage infections enable one to find extremely rare recombinants. The fine structure of genes can therefore be studied by recombination that can resolve mutations just nucleotides apart. In fact, studies of recombination of mutations of the rII gene of bacteriophage T4 gave the first convincing evidence that recombination could take place within a gene, and in fact provided the current conceptual framework, embodied in the *cis-trans* test, that distinguishes mutational sites (positions within genes) from alleles (alternative forms of entire genes).

2. The *rII* system of bacteriophage T4

Fine-structure mapping and deletion mapping in phage T4 are simple extensions of gene mapping that you are familiar with. In two-point crosses, larger distances are signified by greater percentages of recombinants. Recombination in phage T4 is so active that it can be detected at the level of neighboring nucleotides of a gene. Generally, two-point crosses are used because, at the level of individual genes, it is not easy to obtain double- and triple-mutant alleles or to distinguish them from single mutants in crosses.

The *rII* genes of phage T2 are especially useful for fine-structure studies, which were developed by Seymour Benzer in the 1950s and 1960s. The *rII* mutants, lying in two adjacent, complementing genes, differ from the wild type (r^+) both in plaque-size (they are larger; r stands for rapid lysis) and in host range (they will infect *E. coli* strain B, but not strain K). The former character makes them very easy to find during mutant hunts. The host-range difference makes it easy to detect extremely rare r^+ intragenic recombinants among the progeny of a cross. If two *rII* mutants are mated by mixed infection in strain B, the number of r^+ progeny can be measured simply by applying a sample of these progeny to strain K. The recombination frequencies that can be detected are on the order of 1×10^{-8} (1 out of 100 million phage particles), owing to the huge progeny yields and the selectivity of the strain K for r^+ recombinants. In such determinations, geneticists assume that an equal number of double-*r* mutants, carrying both mutations borne by the parents, are present but unrecognizable.

In these experiments, low rates of recombination, measured by the appearance of r^+ phages, often could not be distinguished from the low rates of reversion of one or both of the parental phage ($rII \rightarrow r^+$). Certain other peculiarities of the recombination system of phage T4 made the two-point mapping of the *rII* gene less additive and less consistent than the classical mapping of eukaryotic chromosomes. Therefore, the mapping of the gene was extended by the use of deletion mutants. Among the large numbers of *rII* mutants were many that failed to revert to wild type, even when subjected to mutagenesis. The revertible mutants were presumed to be "point" mutants, that is, base substitutions and frameshifts, while the non-revertible mutants were tentatively identified as deletions. Indeed, the presumed deletions were often found to be unable to yield r^+ recombinants with two or more point mutants, and these point mutants were consecutive on the genetic map drawn from two-point crosses (Fig. 18–7).

Crosses of a vast number of deletion mutants with one another, in contrast to crosses of point mutants, yielded a surprisingly unambiguous map of the *rII* gene. Paradoxically, the crosses did not have to be assessed quantitatively, but only by whether they gave r^+ recombinants or not. The principles of the mapping method were these: if two deletion mutants yielded wild-type progeny, they could not be "overlapping." This meant that the mutants together contained all portions of the *rII* gene, and recombination could reconstruct a wild-type r^+ allele from the parts available in the parents. By contrast, deletions that did not yield r^+ recombinants must overlap; that is, although they might not be identical, they were missing some portion of the gene in common. Consider three mutants, A, B and C, mated in the three pairwise combinations. If A × B yields r^+ recombinants, but neither A nor B does so when crossed with C, then a map of the three would look like this:

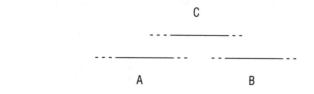

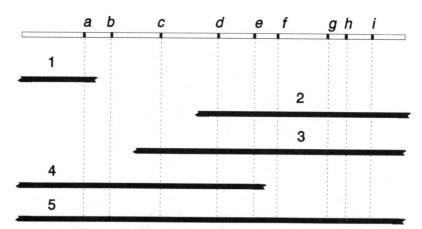

Figure 18–7 Deletion mapping in phage T4. The top bar (open) is the normal DNA with the positions of mutations *a - i* shown. The numbered deletions (black bars) are shown with jagged ends to indicate their indefinite endpoints. For instance, deletion 3 recombines with *b*, but not with *c*, but the left endpoint of deletion 3 cannot be specified more precisely.

The lines represent the portion of the gene that is *missing,* and the dashed ends of the lines signify the unknown position of the endpoints. More deletions in the same area could refine the map from the minimal map above. Furthermore, the position of any point mutation in the area can be quickly localized with respect to the endpoints of deletions (Fig. 18–7), thereby refining the map.

The number of deletions of different size and position allowed a full mapping of the *rII* genes, yielding a pattern consistent with a one-dimensional map, as expected. Since that time, deletion mapping—requiring only qualitative results from the cross—has been used widely for rigorous mapping on a fine scale. It is, however, limited to systems in which deletions are commonly available among the mutations used.

E. Lysogeny

When a temperate phage such as phage λ injects its DNA into a bacterial cell, either of two possible pathways of further development may ensue (Table 18–2). One is the lytic cycle, which has been described above for virulent phages. Alternatively, the phage DNA may become a prophage by integrating into the chromosome of the host bacterium (or more rarely by becoming a separate, dormant circular plasmid). This process is called **lysogenization;** a bacterium with a latent phage is called a lysogen because it can generate a lytic infection at a later time. The pathway taken by an infecting DNA is determined by a number of physiological factors, which converge in determining the amount of a protein, called the phage repressor, in the cell. The phage repressor protein is the product of a gene of the phage. If its accumulation is great enough, there will be enough molecules to bind with high probability to two sites in the phage DNA. When repressor is bound to these sites, transcription of all other genes of the phage is blocked, and the lytic pathway cannot be taken. Because the accumulation of the repressor is opposed by another protein, also encoded by the phage DNA, the competition between the lytic and lysogenic pathways upon infection of a susceptible cell is decided by factors such as the general health of the cell or the phase of growth of the culture.

The DNA of phage λ, which is linear in the virion, becomes circular after injection. The ability to circularize depends upon the 12-base, single-stranded ends of the phage DNA, which are complementary to one another. The pairing, or "annealing" of these "cohesive" *cos* sites on the ends of the DNA is followed by ligation of the first and last nucleotides of each DNA strand, making a double-stranded circle. When repression is effective, and the lysogenic pathway is embarked upon, a protein encoded by the phage DNA, the integrase, catalyzes the recombination of specific sites on the phage and bacterial DNAs. These are the phage attachment site, *attP,* and an attachment site on the bacterial chromosome, *attB.* As in the previous cases of one circle recombining with another (Fig. 18–3), this leads to integration of the phage DNA as a continuous segment of the bacterial chromosome. As this happens, more phage repressor accumulates, and when integration is complete, the only remaining active phage gene is the repressor gene, whose product maintains the phage in the prophage state. As we saw in the discussion of transduction, repressor molecules can also block the development of superinfecting phage DNA. Therefore lysogenic bacteria are immune to further λ phage infection.

The induction of a prophage in a lysogenic bacterium to undergo the lytic cycle occurs by any means that reduces the amount of the repressor protein. Among these agents are ultraviolet light, which works quite indirectly in destroying the repressor. When repressor is gone, other genes of the prophage become actively transcribed, among them proteins leading to the excision of the prophage from the chromosome of the host. (It is in the excision process that rare, specialized transducing phage DNA molecules arise, as explained above.) Excision of the prophage is followed by events of the lytic cycle that culminate in bacterial cell lysis and the liberation of mature virions.

QUESTIONS

1. What are the steps in bacterial conjugation, and what is the rate-limiting step in the acquisition of a given gene by an F⁻ from an Hfr cell in a laboratory situation?

2. What is the function of selective markers in bacterial recombination? What would happen if they were not used?

3. In undisturbed matings, why are there so many fewer recombinants with late genes than with early genes? Why are there so few Hfrs?

4. The F factor is transferred with high efficiency between F⁺ and F⁻ cells. How would you explain the rare transfer of **chromosomal** genes between F⁺ and F⁻ populations?

5. How does an F⁺ bacterial cell become an Hfr?

6. What is the program of events in the growth cycle of a virulent bacteriophage?

7. Define plasmid, temperate bacteriophage, and virulent bacteriophage.

8. If bacterial genes are cotransduced by a generalized transducing phage, what can you conclude about their location, relative to one another, on the bacterial chromosome?

9. Why is generalized transduction not the best method for initially mapping two formerly unknown mutants relative to one another on the bacterial chromosome?

10. What is an auxotroph? In what way does it differ from a "utilization" mutant?

11. How does transformation with fragments of the bacterial chromosome differ from transformation with plasmids in the fate of the donor DNA?

12. In what ways do transposable elements differ from bacteriophage and plasmids?

13. Compare the gene transfer and recombination mechanisms of bacteria and yeast.

PROBLEMS

1. A streptomycin-sensitive strain of *Streptococcus pneumoniae* that was unable to utilize mannitol or galactose as carbon sources was transformed with DNA from a second strain that was streptomycin-resistant and able to grow on either mannitol or galactose as a sole source of carbon. Both strains could use the sugar glucose as a carbon source for growth. Samples of approximately 10,000 (1×10^4) recipient cells were plated on three media selective for these phenotypes: (i) streptomycin + glucose (Strep + GLU); (ii) galactose only (GAL); and (iii) mannitol only (MAN). Colonies growing on each plate were replica-plated to the other two media. The results were as follows:

Original medium	No. of colonies	Number of colonies on replicas made on		
		Strep + GLU	GAL	MAN
Strep + GLU	95	95	0	11
GAL	26	0	26	0
MAN	80	8	0	80

What do you conclude from these observations?

2.   An *E. coli* mating between Hfr *trp*⁺ *his*⁺ *str-s* and F⁻ *trp*⁻ *his*⁻ *str-r* was allowed to proceed for 30 minutes. The mixture was plated on medium containing either (i) streptomycin + tryptophan, or (ii) streptomycin + histidine. In the first case, replica plating revealed that, of the 104 colonies, 56 were *trp*⁺. In the second case, replica plating showed that 10 of the 70 colonies were *his*⁺. Which of the following is the best representation of the location of the *trp* and *his* genes relative to the origin of transfer ($>$) of the Hfr chromosome? (The $>$ symbol points *away* from the markers that enter behind it. Markers to which it points are among the last to be transferred.)

(a)   - - - - *his* - - - - > - - - - *trp* - - - -

(b)   - - - - *trp* - - - - > - - - - *his* - - - -

(c)   - - - - *trp* - - - - *his* - - - - > - - - -

(d)   - - - - *his* - - - - *trp* - - - - > - - - -

(e)   - - - - > - - - - *trp* - - - - *his* - - - -

3.   Phage P1 transduction of four markers (the wild-type alleles of *a, b, c,* and *d*) showed that the percentages of cotransduction were:

$$a\text{-}b\text{: } 29 \qquad a\text{-}c\text{: } 2 \qquad a\text{-}d\text{: } 5$$
$$b\text{-}c\text{: } 0 \qquad b\text{-}d\text{: } 1 \qquad c\text{-}d\text{: } 50$$

What is the order of the markers?

4.   Construct a map from the following crosses between phages carrying different mutant alleles of the *rII* region of phage T2, *a, b, c,* and *d*. The infections were made by mixed infections in *E. coli* strain B, the progeny being plated on strain K in order to determine the frequency of *r*⁺ recombinants. (The form of the crosses is $r^x \times r^y \rightarrow r^x, r^y. r^{xy}, r^+$.) Give the answer in map units.

$$a \times b\text{: } 0.10\% \ r^+$$
$$a \times c\text{: } 0.05\% \ r^+$$
$$a \times d\text{: } 0.25\% \ r^+$$
$$b \times c\text{: } 0.12\% \ r^+$$
$$b \times d\text{: } 0.32\% \ r^+$$
$$c \times d\text{: } 0.15\% \ r^+$$

5.   Construct a map from the following two-factor crosses of phage T4 *r*II mutants. An unambiguous order may not be possible.

Cross	% recombination
*r*145 × *r*168	0.10
*r*145 × *r*228	0.14
*r*145 × *r*295	0.10
*r*168 × *r*228	0.29
*r*168 × *r*295	0.013
*r*228 × *r*295	0.29

6. Consider the *E. coli* chromosome marked into regions I–V by F factor integration sites in five Hfr strains:

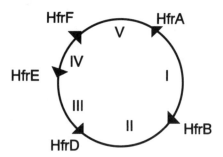

Place the unknown mutations *r, s, t,* and *u* in the proper chromosome regions, using the information below, in which a + sign indicates reasonably frequent transfer, and — represents infrequent or no transfer in Hfr × F⁻ matings.

Mutation	Transfer in matings using				
	HfrA	HfrB	HfrD	HfrE	HfrF
r	—	+	—	+	—
s	+	+	—	—	—
t	—	—	+	+	+
u	—	—	+	—	+

7. In a transduction with the donor *aceF⁺ dhl⁻* and the recipient *aceF⁻ dhl⁺*, selection for *aceF⁺* yielded 88% *dhl⁻* and 12% *dhl⁺*. When the donor was *aceF⁺ leu⁻* and the recipient was *aceF⁻ leu⁺*, selection for *aceF⁺* yielded 34% *leu⁻* and 66% *leu⁺*. Which is closer to *aceF, dhl,* or *leu?*

8. Order the markers *pps, aroD, gurA,* and *man* from the following cotransduction data, giving a qualitative statement of relative distances.

Markers	% cotransduction
aroD - gurA	2
aroD - man	2
aroD - pps	90
gurA - man	90
gurA - pps	2
man - pps	0

9. Two Hfr strains were compared in matings with an F⁻ strain having multiple mutations (in genes *a, b, c, d,* etc.). With HfrA, the markers *a⁺, b⁺,* and *c⁺* were transmitted at 14, 16, and 25 minutes, respectively. With HfrB, markers *a⁺* and *c⁺* were transmitted at 14 and 3 minutes, respectively. What time of entry would you expect for marker *b⁺* in the HfrB strain?

10. An interrupted mating experiment was performed with an Hfr strain that was lysogenic for phage lambda and an F⁻ recipient that was nonlysogenic. It seemed impossible to recover

exconjugants carrying markers that would enter F⁻ recipients after the prophage DNA. Can you suggest an explanation?

11. P1 transduction was done to establish the order of two genes in a group of *leu* (leucine biosynthesis) genes with respect to the flanking *thr* (threonine biosynthesis) gene nearby. The donor was *thr⁺*, the recipient was *thr⁻*, and selection was applied for *thr⁺*. The following crosses were done, with the results (numbers of colonies) as follows:

Donor: *thr⁺ leuX⁺ leuY⁻*	Donor: *thr⁺ leuX⁻ leuY⁺*
Recipient: *thr⁻ leuX⁻ leuY⁺*	Recipient: *thr⁻ leuX⁺ leuY⁻*
4 *thr⁺ leuX⁺ leuY⁺*	17 *thr⁺ leuX⁺ leuY⁺*
30 *thr⁺ leuY⁻ leuY⁺*	40 *thr⁺ leuX⁻ leuY⁺*
20 *thr⁺ leuX⁺ leuY⁻*	30 *thr⁺ leuX⁺ leuY⁻*
16 *thr⁺ leuY⁻ leuY⁻*	3 *thr⁺ leuY⁻ leuY⁻*

What is the order of *leuX* and *leuY* with respect to *thr?* Diagram the recombination events by which the transductants in the left-hand cross are formed.

12. A set of deletions in the r_{II} region of phage T4, when tested pairwise for recombination, gave the following results (+ = r^+ recombinants; 0 = no r^+ recombinants). Make a map consistent with the data.

	a	b	c	d	e	f	g	h
a	0							
b	+	0						
c	0	0	0					
d	+	+	+	0				
e	0	+	0	+	0			
f	0	0	0	0	+	0		
g	+	+	+	+	+	+	0	
h	+	0	0	0	+	0	0	0

Chapter 19

REGULATION OF GENE ACTIVITY

The complexity of growth and metabolism requires precise control of the amount of many proteins. The control is most often exerted by regulatory genes that have evolved to regulate the transcription rates of genes encoding enzymes and proteins of cell structure. Gene expression can also be controlled at other stages. Through these means, organisms have developed great flexibility in responding appropriately to environmental and internal change.

A. Gene Organization, Transcription, and Regulation in Prokaryotes

Organisms control gene activity in response to changing circumstances. Such control conserves energy and resources and directs metabolism efficiently to growth and reproduction. For instance, *E. coli* makes β-galactosidase, an enzyme that breaks down the disaccharide lactose to glucose and galactose, only when lactose is present and when the more easily metabolized glucose is not. This behavior led biochemists to call β-galactosidase and other such enzymes "inducible" or "adaptive" enzymes.

1. Structural and regulatory genes

The unicellular haploid character, simple nutrition, fast growth, and lack of developmental complexity of bacteria and yeast suits them to investigations of gene control. Early studies of mutants unable to turn the synthesis of adaptive enzymes on or off demonstrated the existence of regulatory genes. A distinction was soon made between genes encoding enzymatic and architectural proteins, called structural genes, and regulatory genes, which regulate their expression. The nomenclature is somewhat confusing: regulatory genes encode proteins, and structural genes encode enzymes as well as structural proteins of the cell. The proteins encoded by regulatory genes are called ***trans*-acting factors,** among which we find **activators** and **repressors.** The term *trans*-acting (*trans:* L., "on the other side of") implies that the regulatory gene need not be closely linked to the structural gene that it controls.

Many, but not all, *trans*-acting factors bind the DNA of a structural gene near the transcriptional start site and affect the binding or activity of RNA polymerase. Other *trans*-acting factors interact with those that bind the DNA, but do not themselves do so. We call the binding sites on structural genes ***cis*-acting** regulatory sites (*cis:* L, "on the same side of"). Their action requires that they be attached to the transcribed DNA of the target structural genes.

2. Gene organization

Gene organization contributes to the efficiency of gene expression, and a special vocabulary has developed around it. In prokaryotes, structural genes encoding enzymes of a single biochemical pathway are often lined up consecutively on the chromosome, and they are transcribed into a single mRNA molecule. The word **operon** designates such a segment of DNA, together with its *cis*-acting regulatory sites. One of the latter is the **promoter,** at which RNA polymerase and certain activating proteins bind. These sites all lie prior to the **transcriptional initiation site.** Many operons encode a single protein (and are usually not called operons), while others have a series of two or more coding regions, each called a **cistron.** The latter type of operon is therefore called **polycistronic.** The term cistron has become equivalent to the word *gene* as we have used it in previous chapters, because it is the unit defined as a gene by a *cis-trans* test (*see* Chapter 17). By transcribing all coding regions into one transcript, however, the cell coordinates the levels of related enzymes.

If we visualize gene organization, the commonly used term "5′ end" derives not from the DNA strands of a gene, but from the transcript: the 5′ end is the one transcribed first, RNA being synthesized in the 5′ to 3′ direction. "Upstream" and "downstream" refer to areas of a gene or transcript at the 5′ end and the 3′ end, respectively. Between the 5′ end of a transcript and the beginning of the first coding region lies a leader, or 5′-untranslated region. The coding region or regions of a transcript will be translated into polypeptides, each beginning with the N-terminal amino acid and ending with the C-terminal amino acid. The last codon in the mRNA that is read is followed by one or more nonsense codons that terminate translation. A 3′-untranslated region follows the nonsense codons of the last cistron.

In prokaryotes, ribosomes often begin to translate an mRNA even before the mRNA is finished. Ribosomes bind near the first codon (an AUG) of each coding region, although some of them will remain bound as they terminate one coding region, release the first polypeptide, move in a 3′ direction through a short intercistronic region, and translate the next cistron.

Trans-acting regulatory proteins may act in several ways. At some point in a regulatory system, a regulatory protein binds a small molecule that signals the environmental condition or metabolic status of the cell. These small molecules are called effectors, and they generally initiate or modify a regulatory response. In negative control systems, the *trans*-acting factor is a repressor protein that blocks access of RNA polymerase to the start-site of transcription in response, direct or indirect, to an effector. When this block in transcription is relieved, the operon is said to be derepressed. In positive control systems, an activator protein aids RNA polymerase in binding to the promoter in response to an effector. The gene is then said to be activated. As we will see, a single gene may be subject to both modes of control. Moreover, the DNA-binding ability of a repressor or an activator protein may be either enabled or prevented by an effector.

Many regulatory systems have been studied, and show a great variety of control circuits. Two examples of negative control can be appreciated in the context of metabolism. In the β-galactosidase system, an effector derived from lactose (allolactose) binds to the repressor and prevents the latter from binding to the *lac* operator. In effect, an inducer (lactose) has negated the effect of a repressor. Thus while lactose induces β-galactosidase, we speak of the gene as being derepressed, thereby recognizing the mechanism that prevails during transcription. In the case of biosynthetic enzymes such as those of the arginine pathway, arginine is a negative effector, enabling repression. Only when arginine, the effector, binds to the arginine repressor protein is the repressor able to bind the DNA of the arginine genes and repress them.

If we turn to positive control (which refers to the action of the *trans*-acting protein), the *trans*-acting protein may be effective as an activator when complexed with an effector or when not complexed with an effector, depending upon the system.

3. Control of transcription: the *lac* operon

The *lac* system of *E. coli* was analyzed fully by Jacob and Monod, who provided much of the vocabulary above. This system was in fact the original example of negative control, and it displays positive control as well. The three *lac* genes are organized into an operon (Fig. 19–1), and are transcribed into a single mRNA molecule. The mRNA leader is followed by the coding sequence of the first cistron, *lacZ,* encoding β-galactosidase. This enzyme breaks down lactose to glucose and galactose. The next cistron, *lacY,* encodes the lactose permease, a protein required for lactose uptake into the cell. The third cistron, *lacA,* encodes a transacetylase with an as yet unknown physiological function.

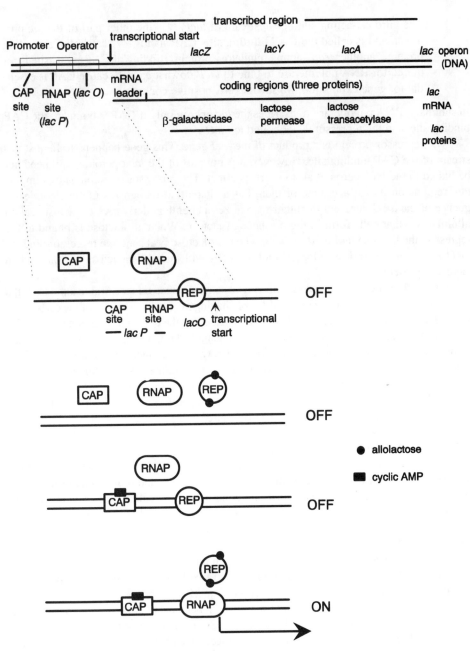

Figure 19–1 The *lac* operon and its control states.

Three *cis*-acting regulatory sites lie upstream of the transcribed region of the *lac* operon. These sites bind particular proteins (Fig. 19–1):

- The **RNA polymerase binding site** is referred to here as *lacP*. In the *lac* operon, the promoter includes this site, together with another site that must be occupied by another protein in order to activate transcription (*see* below).

- Overlapping *lacP* is an **operator** site, *lacO*, which may bind a repressor protein. When the repressor protein is bound to *lacO*, it interferes with RNA polymerase binding at *lacP*. The repressor is a *trans*-acting factor exerting negative control. Operators and promoters may be distinct or overlapping, and their definition is based not on their positions, but on their functions in binding distinct proteins.

- The third *cis*-acting site lies upstream of *lacP*. It is the other part of the *lac* promoter, and is called the **CAP binding site.** CAP stands for "catabolite activating protein," another *trans*-acting regulatory protein. Binding of CAP at this site is required for RNA polymerase to bind to *lacP* downstream. Because CAP is essential for gene expression, we say it exerts positive control.

In summary, RNA polymerase transcribes the *lac* operon only if CAP is bound to the CAP binding site, and the repressor is not bound to *lacO*.

The repressor protein is a product of the *lacI* gene. This gene happens to lie just upstream of the CAP binding site. However, it is not part of the *lac* operon, and it need not be linked to the *lac* operon at all to exert its effect. The repressor protein has two important regions on its surface. One of them has a shape that recognizes the nucleotide sequence of the *lacO* site, and the other recognizes a chemical derivative of lactose, called **allolactose**[1], that cells form in the presence of lactose. When allolactose is bound to the repressor, the latter cannot bind to *lacO;* when allolactose is absent, the repressor binds to *lacO* and represses transcription. Therefore, cells will make β-galactosidase only when lactose is present.

CAP is the product of the *crp* gene, lying at some distance on the chromosome from the *lac* operon. CAP controls many genes of *E. coli,* having evolved as a sensor of the carbon status of the cell. The carbon status of *E. coli* is indicated by the concentration of a small molecule, **cyclic AMP (cAMP).** When cells have adequate carbon, they produce only very small amounts of cAMP. When starved for carbon, however, they produce much more cAMP. CAP binds cAMP, and in doing so, it acquires the ability to bind the CAP binding site in the *lac* promoter. This encourages RNA polymerase to bind at its adjacent site, *lacP* (Fig. 19–1). Thus only when two conditions are met will the *lac* operon be transcribed: (i) if lactose is available, and (ii) if the cell has no other ready source of carbon, such as glucose.

4. Mutants of the *lac* operon

The *lac* system was elucidated by finding *lac* regulatory mutations and determining their physiological effects and genetic relationships. Jacob and Monod found mutants that made large amounts of β-galactosidase under any condition. They called these mutants **constitutive.** Further work yielded other mutants that, although they had a normal β-galactosidase gene, *lacZ,* produced no β-galactosidase. Jacob and Monod called these mutants **uninducible.** They determined the map position of the mutants with conventional crosses, and they determined dominance relationships with merodiploids, using F′ plasmids carrying a copy of the entire *lac* region, including *lacI.* Table 19–1 shows the phenotypes of normal and merodiploid strains carrying mutations at *lacZ, lacO* or *lacI.*

Because the repressor and the *lacO* site are both necessary for repression, disabling either one should lead to high, unregulated rates of β-galactosidase synthesis. Indeed, Jacob and Monod found both sorts of mutant. These, together with genetic data, allowed them to draw important conclusions. First, *lacZ*⁻ and *lacI*⁻ mutants were recessive, and their interaction did not depend upon their *cis-trans* relationship. The observation showed that *lacI*⁺ did not have to be attached to *lacZ*⁺ in order to regulate it. Jacob and Monod inferred that the wild-type alleles encode diffusible proteins. However, the constitutive mutation, *lacO*ᶜ, exerted its effect only when it was physically attached to the *lacZ*⁺ allele, but not when it was separated from it in a suitable merodiploid (Table 19–1). This is consistent with the description of the *lacO* site above. The dependence of the expression upon the arrangement of the genetic material required another term, and Jacob and Monod called mutations such as *lacO*ᶜ "**cis-dominant.**" In general, this term designates sites that—mutant or normal—determine the action of the attached gene, regardless of what homologous sites might lie elsewhere in the cell.

---

[1] This fact was discovered late in the analysis of the *lac* operon. By that time, Jacob and Monod had developed more useful chemical analogues of lactose that bound to the repressor more avidly. These were used for all later studies of β-galactosidase control.

Table 19–1 Phenotypes of *lac* mutants in the *lacI, lacO* and *lacZ* genes, and F′ merodiploids thereof.

Genotype	β-galactosidase on glucose	β-galactosidase on lactose	Comment
Haploid:			
$I^+ O^+ Z^+$	no	yes	wild type
$I^+ O^+ Z^-$	no	no	structural gene mutant
$I^+ O^c Z^+$	yes	yes	constitutive (operator)
$I^- O^+ Z^+$	yes	yes	constitutive (repressor)
$I^s O^+ Z^+$	no	no	super-repressor
$I^{-d} O^+ Z^+$	yes	yes	constitutive (dominant)
Merodiploid			
$F' I^+ O^+ Z^+ / I^- O^+ Z^-$	no	yes	I^+ dominant in *cis*
$F' I^+ O^+ Z^- / I^- O^+ Z^+$	no	yes	I^+ dominant in *trans*
$F' I^+ O^+ Z^- / I^+ O^c Z^+$	yes	yes	Z^- recessive
$F' I^+ O^+ Z^+ / I^+ O^c Z^-$	yes	yes	O^c *cis*-dominant
$F' I^+ O^+ Z^+ / I^+ O^c Z^-$	no	yes	O^c not *trans*-dominant
$F' I^s O^+ Z^+ / I^+ O^+ Z^+$	no	no	I^s product superrepresses; dominant
$F' I^s O^+ Z^- / I^+ O^+ Z^+$	no	no	I^s product superrepresses; dom. in *trans*
$F' I^s O^+ Z^- / I^+ O^c Z^+$	yes	yes	O^c immune to I^s product
$F' I^{-d} O^+ Z^- / I^+ O^+ Z^+$	yes	yes	I^{-d} product blocks action of I^+ product

Two unusual alleles of the *lacI* gene were recovered. One, *lacI*^{s}, encoded a "super-repressor," which was later shown directly to be a form of the repressor unable to bind allo-lactose. Because of this, the mutant repressor remained bound to the *lacO* site and exerted repression regardless of whether lactose was present or not. The allele was dominant. The other unusual mutation, *lacI*$^{-d}$, encoded an ineffective (non-repressing) form of the repressor that behaved peculiarly when tested in a merodiploid. Biochemical investigations showed that the repressor was made up of four identical subunits (polypeptides). The repressor encoded by the *lacI*$^{-d}$ allele had the unusual ability to disrupt the multimeric tetramer even if it was only one of the four subunits. Therefore, this mutation was dominant: because in the merodiploid, there are so few normal tetramers (1/16 of the total if random associations of equally frequent polypeptides are present), repression is ineffective.

Genetic analysis of the *lac* operon also revealed mutations of *lacP* and the CAP binding site. You should be able to predict the phenotypes and dominance relationships of mutations of these sites, and of *crp* mutations affecting the ability of CAP to bind the lac promoter.

5. Attenuation in the *trp* operon of *E. coli*

Another mechanism of transcriptional control, called **attenuation,** acts after transcription begins. In prokaryotes, translation of an mRNA begins before the mRNA molecule is completed: ribosomes are often found on the mRNA just behind RNA polymerase. (This cannot happen in eukaryotes, because transcription occurs in the nucleus and translation occurs in the cytoplasm.) Attenuation is a mechanism controlling the termination of transcription through the translational activity of ribosomes on the mRNA just behind RNA polymerase.

The *trp* operon encodes five enzymes of tryptophan biosynthesis (Fig. 19–2). A nucleotide sequence called the *att* (attenuator) site lies in the leader of the mRNA transcript. The *trp att* site, when transcribed as part of the *trp* mRNA, is a rarity in having a series of 14 translatable codons, called an **upstream open reading frame (uORF).** As RNA polymerase transcribes the *att* site, ribosomes initiate translation of the uORF of the *att* site of the mRNA as it appears behind RNA polymerase. Significantly, the uORF has two consecutive tryptophan codons (UGG).

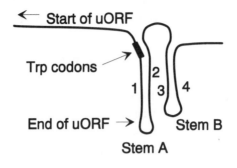

5′ end of
trp mRNA
after removal of
all proteins: pairing
of 1+2 (Stem A) and
of 3+4 (Stem B)

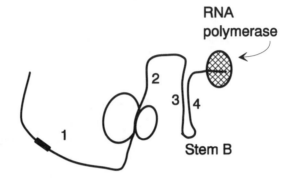

Tryptophan-
sufficient cells:
ribosome blocks
formation of Stem
A; Stem B forms.
Stem B blocks further
progress of RNAP

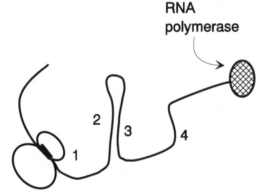

Tryptophan-
limited cells:
ribosomes stall at
trp codons; segment
2 is free to pair with 3.
Stem B cannot form,
and RNAP continues.

Figure 19–2   Attenuation in the *trp* operon of *E. coli.* Base-pairing options 1-2 (stem A), 3-4 (stem B), and 2-3 are determined by the availability of tryptophanyl-tRNA, which in turn modifies the ability of ribosomes to traverse the entire uORF.

The key feature of attenuation is the use of a mechanism that often terminates transcription. At the end of many genes is a sequence that, when transcribed, through internal base pairing, folds the mRNA into a shape that acts as a transcription-termination signal. This sequence, as it emerges from RNA polymerase, but still in contact with it, brings transcription to an end. This mechanism has been adopted in the *att* site as a regulatory feature.

The *att* segment of the mRNA, if fully formed and not bound by any proteins, will adopt a two-stem structure by internal base-pairing of segments 1 + 2 (stem A) and 3 + 4 (stem B) (Fig. 19–2). Stem B is a transcription termination signal. In the discussion that follows, we will find that an alternative pairing arrangement can be adopted, such that segment 2 pairs with segment 3 and preempts the formation of stem B. Tryptophan indirectly modulates the choice of forming the 2 + 3 pair vs stem B, and thus the continuation of transcription of the operon.

If tryptophan is abundant, ribosomes translate the uORF, including the tryptophan codons, of the *att* site easily. Their progress down the uORF engages the nucleotides of segment 1 and segment 2 of the *att* site. As transcription continues, segment 3 appears, but cannot pair with segment 2 because ribosomes continue to occupy it. Therefore, when segment 4 appears, the transcription-termination structure (stem B) forms between the uORF and RNA polymerase. Transcription stops before RNA polymerase transcribes the first *trp* coding region. The cell therefore makes no tryptophan enzymes.

In cells starving for tryptophan, however, there is little trp-tRNA, which is required to "read" the tryptophan codons in the uORF. Therefore, ribosomes will "stall" at these tryptophan codons. The ribosomes therefore do not occupy the nucleotides of segment 2 that can base-pair with segment 3. As segment 3 appears behind RNA polymerase, the alternative pairing arrangement (2 + 3) forms, pre-empting segment 3 from forming stem B with segment 4. Without the terminator-stem B, transcription continues through the five *trp* cistrons, and the cell makes tryptophan-biosynthetic enzymes.

B. Gene Organization and Regulation in Eukaryotes

1. Scattered genes

Eukaryotes differ significantly from prokaryotes in certain attributes of gene organization and regulation. Grouping of genes of related function is much less common in eukaryotes, at least for genes controlling metabolism and growth. With few exceptions, this is correlated with a **monocistronic** character of eukaryotic mRNAs: each mRNA encodes only one protein. This property reflects a difference in the way in which eukaryotic and prokaryotic ribosomes read mRNAs. The small subunit of a prokaryotic ribosome will bind to the beginning of a coding region, wherever it might be in an mRNA. Moreover, when it comes to the end of a coding region, it remains associated with the mRNA and can potentially encounter another coding region. In the case of eukaryotic ribosomes, however, the small subunit binds first to the 5' end of an mRNA (the m^7G cap), and then "scans" the leader until it reaches the first AUG codon. Then, after the engagement the large ribosomal subunit, the ribosome begins translation. At the end of the coding region, the subunits dissociate and cannot read another coding region downstream. Hence, polycistronic operons would not be expected in eukaryotes.

The tendency of related genes in eukaryotes to be scattered allows the linkage relationships of such sets to vary by translocation during evolution without serious effects on function. However, sophisticated mechanisms of coordinating gene activity have developed in eukaryotes.

2. Regulation of eukaryotic genes

The regions lying 5' to the transcribed regions of eukaryotic genes are grossly similar to those of prokaryotes, but they are more complex. Promoters are longer, and consist of binding sites for several non-regulatory proteins in addition to RNA polymerase and regulatory proteins. The sites, named for their most common sequence, are the "CCAAT box" and the "TATA box," each of which binds a special protein before RNA polymerase will bind in the same area. As a result, the so-called transcription complex, which includes RNA polymerase, the CCAAT-binding protein, the TATA binding protein, other proteins of the complex, together with one or more regulatory proteins, must assemble at the promoter before transcription begins.

Some eukaryotic genes have *cis*-acting sites called repression sequences, operators that are bound by negative, *trans*-acting proteins. These sites are functional equivalents of bacterial operator sequences. More frequently, however, *trans*-acting factors control genes positively by binding *cis*-acting elements in the promoter or in more distant sites we call **upstream activation sites,** distinct from the promoter. These sites are rather different from the positive control elements of prokaryotes, because they can be at some distance upstream of the gene they control. They may encourage the binding of RNA polymerase or other *trans*-acting factors. Some activating sequences called enhancers are even freer in their location: they may be either upstream or downstream of the gene they influence, and may act whether they point to or away from the gene in question. The distinction between upstream activating sequences and enhancers is hard to make unless tests of varying the orientation and position have been done. The length of the DNA over which these interactions takes place requires loops to form in the DNA so that the binding proteins can interact. Indeed, many *trans*-acting factors impart bends in the DNA without which the interactions would not take place.

Higher eukaryotes, owing to their complexity, have a number of additional controlling mechanisms whose details are beyond the scope of this treatment. However, one of them is widespread in vertebrate and higher plant genomes; namely, a heavy **methylation** of DNA of genes that are silenced during differentiation. Methylation of DNA is confined largely to one of the carbon atoms of cytosine. The mechanism by which this modification of DNA is initiated and controlled during development is not clear, but it is easily maintained during replication of cells. The methylation of DNA is really a confirmatory, and more permanent step in the process of repressing genes, rather than the initial step, which involves *trans*-acting factors. Methylated DNA is found in great abundance in heterochromatic, transcriptionally inactive chromatin. The formation and spacing of nucleosomes (*see* Chapter 2) may also modulate gene expression as well.

A number of "fusion genes" have evolved in eukaryotes, by which related enzymes have been brought together through selection of random translocations and other chromosomal aberrations. Well-evolved fused genes produce multi-enzyme proteins, single polypeptide chains that are translated from a long mRNA with a single translation-initiation step. Superficially, these genes resemble polycistronic operons, but their translation products are single polypeptides, not several. The protein segments corresponding to separate enzymes may remain fused as a multi-enzyme polypeptide, or the multienzyme polypeptide may be cut into smaller proteins after translation. Either way, fused genes present eukaryotes with an effective means of coordinating the expression of many proteins, comparable to the operon organization in prokaryotes.

C. Complex Regulation in Pro- and Eukaryotes

Prokaryotic and eukaryotic organisms share another means of coordinating activities of genes needed at the same time, namely, common regulatory sites such as upstream activation sites, enhancers, and operators (all called consensus sites in this context) at the 5′ ends of the genes of such repertories. Eukaryotes do not have multicistronic operons, and fused genes are somewhat exceptional. Therefore, they rely heavily on regulatory coordination by *cis*-acting consensus sites. Thus we find upstream sites responsive, for instance, to CAP (catabolite activation protein), that signals carbon deprivation. Pro- and eukaryotes have other *trans*-acting factors responding to starvation for carbon, nitrogen, sulfur, phosphorus, or to amino acid imbalance, which activate or relieve repression of appropriate genes. Other *trans*-acting factors activate or discontinue repression of genes needed after heat shock, osmotic shock, inhibition of protein synthesis, and ultraviolet irradiation (the SOS response). We call these responses global regulatory responses, because of the general nature of the inducing agent, and because so many genes are involved.

We have noted above that not all *trans*-acting factors bind DNA, but often bind those that do. We see, particularly in eukaryotes, an extension of this pattern into a regulatory cascade that culminates in the expression of a gene repertory. In a regulatory cascade, one regulatory protein may bind an environmental molecular signal and then enter the nucleus. There, a second regulatory protein may bind the first. The complex of proteins may then activate the synthesis of a positive regulatory protein that, in turn, activates one or more structural genes. Alternatively, the complex may interfere with repression by a repressor protein. A number of such cascades have been studied in simple eukaryotes such as yeast, using mutants for each step to define the cascade, much like auxotrophic mutants were analyzed to define metabolic pathways.

Even the simplest control mechanisms give great flexibility to organisms in their response to the environment. Many individual genes, if they are parts of several different repertories, have binding sites for a variety of regulatory proteins. Therefore, the array of regulatory sites gives a **combinatorial** character to regulation, and allows cells to produce appropriate and different constellations of enzymes in different conditions. The differentiation of complex higher organisms builds heavily upon this feature of gene expression.

Still other control mechanisms are well known. It suffices here to name the steps at which these other control systems exert their effects. In translational regulation, translation of mRNA, rather than transcription of the gene, is the key step controlling the amount of a protein. In some translational mechanisms, the sequence or folding of the mRNA leader forms a signal to which *trans*-acting factors respond. In some mRNA leaders of eukaryotes, uORFs that interfere with the scanning of mRNA leaders by ribosomes have evolved as a regulatory mechanism. Regulation of gene expression through mRNA stability is often seen in eukaryotes, and both pro- and eukaryotes regulate many metabolic activities by regulating the stability of proteins. Finally, a major form of control is found at the level of enzyme activity: many biochemical pathways include **feedback inhibition,** a mechanism by which the endproduct of the pathway inhibits one of the earliest enzymes in the pathway. The endproduct binds to a special allosteric (Gr., "other shape") site of the early enzyme and changes its shape so that it catalyzes the reaction less efficiently or not at all. This allosteric phenomenon is similar to the binding of allolactose to the *lac* repressor and cAMP to the CAP protein.

QUESTIONS

1. What is the difference between an operator, an attenuator, and an enhancer?

2. What is the difference between an activator protein and a repressor protein?

3. What is the difference between an operator and a repressor?

4. What would the phenotype of a *lac* mutation defective in the CAP binding site be? What would its dominance relationship be to its wild-type counterpart?

5. *Cis*-dominant, uninducible mutations in the 5′ region of the *lac* gene prior to the transcribed region were found. Can you think of what they might be?

6. In *E. coli,* the study of the *lac* operon was greatly aided by the availability of F′ factors. What indispensable experiments did F′ factors allow Jacob and Monod to do?

7. Why is the *E. coli lac* operon uninducible in the presence of glucose?

8. Positive and negative control are different ways of controlling the activity of a gene. What is the difference, and can one gene be subject to both modes of control?

9. Why are mutations of the *lac* operator often called *cis*-dominant? How can a mutation in the *lacI* gene be *trans*-dominant?

10. How does an attenuator exert its regulatory effect?

11.   The rate of certain metabolic activities in eukaryotes vary greatly according to environmental conditions, but the rate of transcription of the relevant genes remains constant. Suggest several levels and mechanisms by which this might happen.

12.   A mutation imparting constitutive synthesis of an enzyme of arginine synthesis in *Neurospora* was found; the enzyme is normally repressible by arginine. (i) What two regulatory mutations might cause this? (ii) What kind of mutation in an enzyme of arginine biosynthesis might cause this?

Chapter 20

GENETIC ENGINEERING

The field of recombinant DNA and genetic engineering relies on enzymes and techniques that permit the specific cutting, splicing, and sequencing of DNA molecules, the recognition of the recombinant products, and the introduction of recombinant molecules into almost any kind of cell. The field has become one of the most vigorous in science, owing to the analytical power it gives us to study gene structure and expression, to the ability it gives us to make many biological products in lower organisms, and the possibility of directly changing the genotypes of organisms of commercial interest.

A. Restriction Enzymes

The study of genes themselves, as opposed to their effects, became possible with the advent of **restriction enzymes.** These enzymes are **endonucleases** found in bacteria as defenses against foreign DNA (mainly from phages) that enter the cell. Each restriction enzyme recognizes a short DNA sequence, or restriction site, and cuts the DNA at or near these points. This inactivates the foreign DNA. While the same sites are found in the bacterial DNA, those of the bacterium are **methylated** on one of their nucleotides, which protects the sites from attack by the restriction enzyme. Thus the presence of a restriction enzyme and modification of DNA by methylation are coupled: for every endonuclease, there is a methylase that recognizes the same sequence. In some cases, the same protein catalyzes both chemical reactions.

The restriction endonucleases most useful in genetic engineering, as noted in Chapter 1, recognize specific sequences of four to ten nucleotide pairs in DNA molecules and cut both chains at those sites. A large number of these enzymes, isolated from bacterial species, are now used to analyze and manipulate DNA of plasmids, phage, bacteria, and eukaryotic genomes. Their names reflect the origin of the enzymes; the enzyme *Eco*RI is derived from *E. coli,* and cuts any DNA at the nucleotide sequence GAATTC; the enzyme *Bam*HI comes from *Bacillus amyloliquifaciens* H and cuts the sequence GGATCC, and so forth.

The nucleotide sequence recognized by most of these enzymes is **palindromic;** that is, it has the same sequence, 5′ to 3′, on each strand. An example is the site for *Eco*RI:

5′ GAATTC 3′
3′ CTTAAG 5′

Many restriction enzymes make **staggered cuts** in the DNA such that there is an identical overhang of 2–4 bases at the cut ends. The utility of staggered cuts is that molecules from different DNAs can be mixed together and religated with DNA ligase to form a recombinant DNA molecule, owing to the complementarity of the overhangs of the two partners. Other restriction enzymes make **blunt cuts;** that is, a cut at the same point in each strand.

Blunt-cut pieces of DNA may be also ligated together with somewhat lower efficiency, and there is no specificity to the pairing of the ends that are ligated, as there is with ends with overhangs. Because endonuclease sites appear frequently in any DNA, molecular geneticists have become quite versatile in making recombinant molecules. We can in fact calculate the average frequency of a particular sequence such as CATG (nucleotides of one strand suffice to represent the sequence) by the multiplication rule, assuming equal numbers of each nucleotide. There is a choice of four nucleotides at each position; therefore, CATG will appear once in every 256 (4^4) nucleotides. The longer sequence AAGCTT will occur once every 4096 nucleotides *on the average* by the same reasoning. Because these lengths of DNA cover within-gene distances to those extending many genes, one can prepare many specific recombinant molecules by using a combination of different enzymes.

Early restriction enzyme analysis began with simple molecules, such as those of a pure preparation of a bacterial plasmid or a phage. The product of such an analysis is a **restriction map,** which depicts the position of different restriction sites within the molecule. Consider a circular molecule 10 **kilobases (kb)** long, with the sites distributed as in Figure 20–1. If the DNA is cut with the enzyme *Eco*RI, two pieces of DNA will be released, 3.6 and 6.4 kb long. Because they differ in size, they may be separated by **electrophoresis,** a

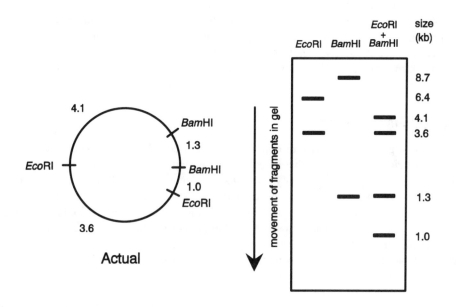

Inferred from gel...

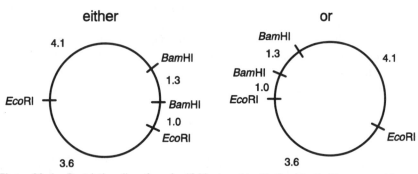

Figure 20–1   Restriction digestion of a 10 kb plasmid with *Bam*HI, *Eco*RI, or a combination of the two, followed by separation of fragments on an agarose gel. The pattern on the gel allows two possible interpretations, which must be distinguished by use of other enzymes or techniques.

technique in which the sample is put at one end of a gel made of agarose (a refined form of agar), and the gel is then subjected to an electric field. DNA, negatively charged owing to its phosphate groups, moves through agarose to the positive pole, with small fragments moving faster than large fragments. The differences in mobility reflect the ease with which small pieces of DNA move through the interstices of the gel, compared to larger pieces. With size standards, the sizes of the pieces can be determined. Once separated, DNA fragments may be visualized in ultraviolet light if the gel is stained with **ethidium bromide,** an intercalating, fluorescent dye.

In restriction mapping, we benefit from knowing that the sum of the sizes of the fragments equals the size of the original DNA molecule. Because circular molecules have different mobility than linear molecules of the same length, circular molecules must be rendered linear by an enzyme that cuts the circle in one place if their sizes are to be measured accurately. However, if *Eco*RI yields 6.4 and 3.6 kb fragments (Fig. 20–1) from a circular molecule, we can calculate the size of the uncut molecule as 10 kb.

If the DNA in our example is cut with another enzyme, *Bam*HI, for which there are two other sites, we will see two pieces again (Fig. 20–1). One piece is 1.3 kb, and the other 8.7 kb in size. A "double digest," made with both enzymes simultaneously, yields four pieces. In our example, the smaller *Eco*RI fragment (3.6 kb) is unaffected by *Bam*HI, and the small piece in the *Bam*HI digest (1.3 kb) is untouched by *Eco*RI. This means that both *Bam*HI sites are in the larger *Eco*RI fragment. We can infer that the smallest piece and the largest piece in the double digest both have a *Bam*HI site at one end and an *Eco*RI site at the other. At this point, we could draw two potential maps of the plasmid with the two enzymes' sites (Fig. 20–1). The two maps differ in the end of the 6.4 kb fragment at which the 1.3 kb fragment lies. Another enzyme or another strategy is necessary to decide the question.

Many other strategies of mapping linear and circular molecules are now in wide use. Among the methods used are the radioactive labelling of the ends of a large linear molecule, which identifies end-fragments in a restriction digest, and partial digestions in which one can visualize intermediates in the digestion process unique to certain possible maps.

B. Vectors and Cloning of DNA Fragments

With a restriction map, we may go on to **clone** specific fragments of DNA. To do so, the fragment of DNA must be "inserted" into a DNA that will replicate in *E. coli* after it enters the cell via transformation. Many small plasmids (2–5 kb) have been engineered as **vectors** (carriers of the DNA of interest) to serve this purpose. These plasmids have three main elements in common: (i) An origin of replication, which all plasmids normally have; (ii) A gene that endows the transformed bacterium with resistance to an antibiotic, such as ampicillin, kanamycin, or tetracycline. This gene allows selection of bacteria that have been transformed from the large number of bacteria that remain untransformed. (iii) A segment of DNA with a number of different restriction sites, all unique within the plasmid DNA. These segments, called **multiple cloning sites,** are now made by chemical means and are introduced into the vectors by ligation.

Most small plasmids are represented by many copies in each bacterial cell. They also have a supercoiled form, which gives these small molecules properties that make them very easy to separate from the main DNA of the bacterium. These two attributes make the isolation of large amounts of plasmid an extremely efficient task, and they underlie the routine character of cloning genes in this type of vector. Other vectors are used; notable among them are derivatives of phage λ which can be induced to undergo the lytic cycle from the prophage state at will. After induction, large amounts of phage or phage DNA can be isolated with ease.

With simple DNA preparations, restriction fragments of a DNA of interest may be isolated and purified from the agarose gel in which they are separated. The nature of the cut ends of the pieces are known, guiding the investigator who wishes to clone fragments to cut a suitable plasmid in its multiple cloning site with the same enzyme(s) used to cut the

DNA of interest. By simply incubating the fragment of interest and the cut plasmid together with DNA ligase, many circular molecules of plasmid with a DNA insert appear. Only circular molecules will propagate in *E. coli,* and therefore all unligated plasmids and free fragments of DNA will be excluded. Some transformants will contain plasmids with insert and some will contain plasmids that have simply been restored to circular form. Each colony of transformed bacteria will contain only one type of plasmid. Many methods have developed with which one can distinguish the two types. After isolating a colony (a clone) of the desired type, the bacteria can be grown in large numbers, the plasmids isolated, and the DNA insert of interest isolated by restriction enzyme digestion. By this and comparable methods, pure fragments of DNA can theoretically be obtained. Practical, detailed manipulation of DNA is done in the size range of 0.5 to 15 kb.

C. Sequencing DNA Fragments

The nucleotide sequence of DNA fragments can now be determined by a variety of methods, many of them highly automated. This is the ultimate, detailed knowledge of the genetic material, though it reveals nothing *per se* of its function. Once a DNA sequence is known and a coding region is identified, the latter may be "translated" into the amino acid sequence that it encodes. By comparison with other sequences, the protein may be identified. DNA sequences are the basis of the knowledge of how mutations—introduced artificially or found in mutants—affect processes at the phenotypic level. DNA sequencing techniques are beyond the scope of this chapter, but they are routinely used in all molecular biological laboratories.

D. Detection and Isolation of DNA Fragments from Complex Genomes

While simple plasmids and simple DNA preparations can be manipulated with ease, the complexity of the DNA of most organisms, even prokaryotes, precludes isolating individual fragments directly from gels, even if one knew the size of the fragment one was looking for. Consider the genome of *E. coli,* with approximately 4,200,000 nucleotide pairs (4200 kb). If *E. coli* DNA were cut with a restriction enzyme with a 6-nucleotide recognition site (a "six-cutter") and then subjected to electrophoresis, approximately 1000 fragments would be spread throughout the gel, ranging from very small (<100 nucleotide pairs) to rather large (over 20 kb). The overlap in the positions of these fragments on an agarose gel is such that any small part of the gel would contain a heterogeneous population of DNAs. In fact, restriction analysis of even the simplest DNA does not alone tell us what genes or parts thereof the fragments might contain. Therefore, a great deal of the technology of gene cloning has been devoted to detection and efficient isolation of genes of interest.

1. Probing gels

If one already has a DNA of interest in hand and wishes to detect the same or a homologous sequence in a gel after restriction enzyme digestion, one can prepare a **Southern blot** (named for scientist E. Southern). The gel is placed on a nylon membrane that will trap and bind DNA avidly. The DNA in the gel is transferred by capillary action to the membrane; the membrane is dried; and it is then **probed,** using a radioactively labelled preparation of the DNA one already has.

The technique depends upon the property of single-stranded DNA molecules to **hybridize** with complementary sequences, also single-stranded, by base-pairing. This is a remarkably specific process. When heated, the two strands of even very long DNA molecules unwind and separate from one another due to the breaking of hydrogen bonds. If cooled quickly, the strands will not have a chance to recognize their complements in a complex mixture. Instead, they collapse into a "random coil." However, if the mixture is cooled very slowly, or if it is given time at a relatively high temperature, complementary strands will find one another, hybridize, and be restored to the double-helical form. The reaction is faster if one of the complements is very short, or if one or both complements is present in high concentration.

This principle is used in probing Southern blots. Short pieces of the DNA already available are used as templates in a DNA polymerase reaction during which radioactive nucleotides (labelled with radioactive phosphorus, ^{32}P) are incorporated into replicas. After blocking the surface of the nylon membrane so that it will no longer bind additional DNA non-specifically, it is then exposed to the labelled DNA in relatively high concentration. The radioactive probe hybridizes with the complementary piece or pieces of DNA on the blot. The blot is washed free of excess radioactivity, laid on an X-ray film, and after a suitable time, the film is developed, revealing the positions of DNA complementary to the probe.

Probing gels permits investigators to visualize DNA fragments in extremely complex mixtures, such as the DNA of the human genome. Therefore, a specific sequence can be detected and its size determined simultaneously, unlike the ethidium bromide method described above. However, the method depends upon having a DNA of interest already, and it does not offer a way of isolating a DNA of interest from the gel.

In order to isolate a fragment from a complex restriction digest, the mixture must be cloned *en masse*, a technique called **shotgun cloning.** Here, a complex DNA is digested with a restriction enzyme and incubated with many molecules of a cut plasmid and DNA ligase. The products, many plasmids with different inserts, are then used to transform *E. coli*. Many individual transformants are obtained, each with a different plasmid containing a different cloned piece of DNA. When spread on an agar plate, the transformants represent a **library** (or bank) of different clones. The next task is to isolate a clone of interest. The DNA of bacterial colonies, after they are lysed on a nylon membrane, will bind and offer itself for probing by the method used in the Southern blotting technique. The probe will allow the detection and isolation of the clone of interest. Again, however, one must have a nucleic acid of the appropriate sequence already.

2. Detection of clones of interest in a library

Many additional methods have been devised to detect particular genes in a complex library. Only several of the common methods will be detailed here in order to illustrate the radically different approaches.

a. cDNA.   Certain RNA viruses transcribe their RNA into DNA upon infection of cells, using an enzyme, reverse transcriptase, that they encode. (The RNA of these viruses also serves as an mRNA for two other proteins.) Once transcribed into DNA, the virus may insert itself into the DNA of the host. Reverse transcriptase has found wide use in making libraries of **cDNAs,** DNA copied from mRNA populations. With cDNAs, geneticists can then make clones as described above.

In pursuit of the gene for a particular protein, molecular biologists often make antibodies to the protein. When cDNAs are cloned in a special λ prophage form that is transcribed and translated in *E. coli* upon induction of the prophage, all or part of the protein encoded by the cDNA appears in the lysed bacterial colonies. With a membrane, replicas of the lysed colonies are captured, and the proteins that adhere to the membrane are probed with antibodies specific for the protein of interest. The antigen-antibody complex is detected by a special, radioactively labelled protein (protein A from the bacterium *Staphylococcus aureus*) that binds to antibody molecules. Phage from the appropriate colony are isolated from the original plate, and the cDNA of interest thereby becomes available for study. The rationale is very similar to probing the nucleic acid with a radioactively labelled DNA. With this method, cDNAs for many genes have been isolated. This approach to eukaryotic genes has the benefit that cDNAs—made from spliced, mature mRNA molecules—have no introns. That is why they are transcribed and translated in bacteria, which have no interrupted genes. Once a cDNA is available, it may be used as a probe in searching for the gene itself in a genomic library of the same organism. The net result is that after sequencing the cDNA and the genomic DNA, the investigator knows not only the protein coding portions of the gene, but the position, length, and sequence of the introns as well.

b. Complementation.   DNA-mediated transformation has become possible in most genetically well-known organisms. This offers the opportunity to introduce any piece of DNA into cells. In many cases, the cell inserts some of the DNA, usually at random locations, into its chromosomal DNA. In some cells, transformants can be recognized by a selectable or visible marker carried by the vector DNA. One may then ask whether the DNA fragment from the library, also carried in by the vector, is expressed in the transformed cell.

A common method of determining whether a DNA is expressed in bacterial and fungal cells depends upon having a mutant, such as an auxotroph, for the gene in question. Indeed, the gene introduced by transformation often restores the phenotype to wild type, thereby making the transformed cell uniquely selectable over all other transformants. Consider arg^- cells, carrying a mutation of the arg-2 gene, which you have transformed with a plasmid genomic library made from wild-type cells. If the transformants are plated out on minimal medium, cells transformed specifically with the arg-2^+ gene will grow; others will not. While it may be difficult to retrieve the plasmid or the gene from the transformed cell, it is nevertheless possible to use the mutant cells as testers, by which large groups of plasmids are tested for the presence of the arg-2^+ gene. By successively subdividing the population of plasmid clones, one can finally identify individual plasmids that complement the arg-2^- mutant. In the case of yeast, plasmids have been developed that replicate both in yeast and bacterial cells. These plasmids can be retrieved from the transformed cells because they do not integrate into the chromosomes. All one needs to do is to take the plasmids from the successfully transformed cells and retransform *E. coli* with them. From this step, one obtains a pure clone. Many genes have been isolated by the use of these "shuttle vectors."

c. Homologous probes and oligonucleotides.   The amino acid sequences of many polypeptides have been determined. When sequences of the same protein from a number of species are compared, one finds considerable variation in some parts of the polypeptide, but other parts are much more similar. The latter, **conserved regions** are constrained to remain similar over evolutionary time, because natural selection preserves the function of the protein. The gene for one species can be used to probe a DNA library of another species in search of the homologous protein. This technique often fails, owing to the increasing divergence of amino acid and nucleotide sequences over evolutionary time; even conserved regions may differ at the level of nucleotides because synonymous codons have been substituted for many amino acids.

A more direct approach is possible if the amino acid sequence of parts of the protein encoded by the desired gene has been determined. Here, it is not necessary to assume homology; it is possible to synthesize an **oligonucleotide** corresponding to the codons for the known amino acids. Actually, this must be a mixed oligonucleotide, owing to the synonymy of many codons. A sequence of 6–8 amino acids can be found in a protein in which there are relatively few synonymous codons. By chemically labelling this sequence with radioactive phosphate, one has a serviceable probe.

d. Polymerase chain reaction (PCR).   This new methodology has widespread use in many fields. The method allows one, potentially, to **amplify** DNA of any sequence from any source, even if it is only a tiny fraction of the DNA sample of interest. The method involves the use of oligonucleotides as primers for the synthesis of complementary strands, using the "target" DNA as a template and a special, heat-stable DNA polymerase. The oligonucleotides may be a mixed oligonucleotide, as described above, or it may be a short natural sequence that will hybridize with the target. A target DNA is heated in order to separate the strands, and then it is allowed to hybridize with oligonucleotide primers 20–30 bases long (or longer). The primers are designed so that one is homologous to the 5' end of the sequence of interest, and the other is homologous to the 3' end of the gene, but actually complementary to the other strand. Many copies of the primers are added initially to the reaction. Because their 3'OH ends are pointing toward one another, DNA polymerase will make the complements of both separated strands, starting with the primer as a defined, 5' end. This will yield two molecules for each one in the target DNA (Fig. 20–2). The chain reaction aspect of the technique follows. The preparation is heated again to 95° C to separate the strands of the two new molecules, then cooled to let further molecules of primer

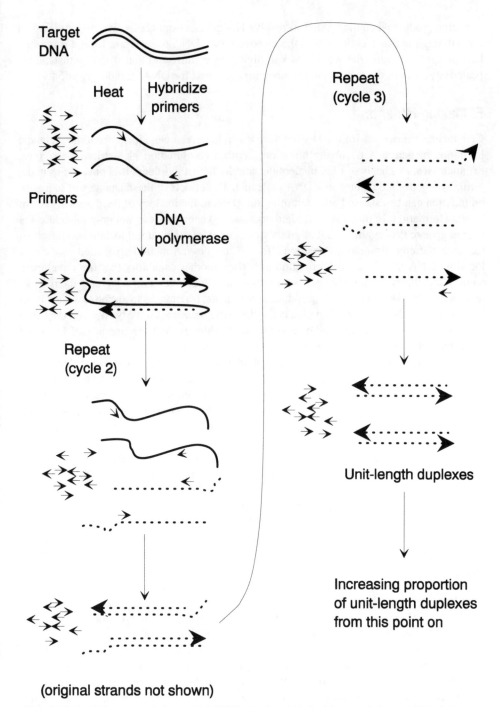

(original strands not shown)

Figure 20–2 Polymerase chain reaction (PCR), starting with double-stranded target DNA. Primers (small arrows) designed to hybridize with specific sequences in the target prime heat-stable DNA polymerase to make new complements of the target DNA. These new strands are of indefinite length. The new strands are then copied in the second cycle, in which strands of unit length appear. Duplexes of unit length appear in the third cycle and become a predominant product thereafter.

hybridize. The DNA polymerase then makes more DNA, yielding four double strands. Two of these are of defined length, and they will become templates in the next cycle for synthesis of more of the same (Fig. 20–2). This continues indefinitely with an increasing number of copies of the DNA of a length defined by the distance between the primers in the target DNA. After 25 cycles, each only a few minutes long, each original molecule is represented by over 30 million copies of defined length. The product can easily be seen as a specific product when the mixture is resolved by gel electrophoresis. The major technical breakthrough in this technique was the discovery of a DNA polymerase from a primitive, hot-springs bacterium that would work at high temperature, and that would withstand repeated exposure to the even higher temperature required for DNA strand separation.

E. Making Mutants

Genetic engineering, as implied by the term itself, has freed geneticists from the technical limitation of naturally occurring mutations. With a combination of restriction enzymes, exonucleases, synthesis of oligonucleotides and PCR methodology, it is possible to make virtually any desired change in a DNA molecule. The effects of these changes in sequence on function can be assessed after returning the DNA to the nucleus of the parent organism by transformation. In many well-studied organisms, geneticists can not only introduce the altered gene to the organism, but often they can target it to the natural location on the chromosome, thereby eliminating "position effects," changes in function associated with a new location. DNA-mediated transformation underlies another capability that has developed, namely, the ability to introduce genes into organisms that do not normally have them. Thus resistance to diseases have been introduced into plants; resistances to antibiotics have been introduced into fungi and mammalian cells; domestic mammals have been given the capability to produce human antibodies and peptide hormones; and the ripening of tomatoes has been reprogrammed for better commercial success.

The most widespread use of "in vitro mutagenesis" is restriction-enzyme deletion of sequences near a coding region to determine the minimal DNA sequence needed for expression and normal regulation of a gene. Once critical sequences have been identified in this way, finer deletions can be introduced by exonuclease action on the DNA, in which the exonuclease chews away nucleotides from one end of the molecule. After cloning a number of the products of the digestion, the sequence of each one can be determined and a set chosen to test for function. Finally, specific nucleotide substitutions can be introduced by PCR or other techniques into small critical regions to determine their tolerance for changes.

In the course of determining the location and nucleotide sequences of critical coding or control regions, geneticists discover the additional benefits of changing them. Most of the mechanistic questions posed by earlier work in genetics and pursued with induced random mutations can now be approached with directed, molecular techniques. Questions ranging from what happens when individual nucleotides are changed to the effects of deletion of entire genes from an organism can now be answered, often very easily. We do no more here than note the direction genetics moves; the field is open and moving fast.

Many goal-oriented changes can be made for both analytical and productive purposes. For instance, removal of a *cis*-acting, negative control site will in many cases greatly increase the rate of production of an enzyme. The same outcome can be achieved by recombining a very active promoter with a coding region for a protein made normally at low levels. In both cases, the effects of overexpression of the gene can be seen, or the enzyme may become easier to purify. Many proteins of pharmaceutical importance are now made in this way, usually using a bacterial host, but eukaryotic production systems are becoming more common.

F. Summary and Conclusion

Only some of the basic areas of genetic engineering have been noted above, mainly to show the logic of its simpler procedures. However, its techniques have allowed the growth of many scientific and commercial enterprises because of the power it gives over biological systems. This power has endowed the field with commercial and ethical problems that were unknown a few years ago and will develop with the benefits, into unforeseen issues in the future.

QUESTIONS

1. What is the advantage of "overhangs" in the making of recombinant DNA molecules?

2. A restriction digest of a 9 kb linear molecule with the enzymes *Pst*I and *Sma*I and a combination yielded fragments of the following sizes: *Pst*I: 3.0 and 6.0; *Sma*I: 4.5, 2.5, and 2.0; *Pst* + *Sma*I: 3.5, 2.5, 2.0, and 1.0. Give the maps compatible with the fragment sizes given.

3. Why do restriction enzymes not destroy the DNA of the cell that makes them?

4. How would you find a cDNA corresponding to an abundant protein that you had purified from mouse liver?

5. You have a histidine-requiring yeast strain with a mutation in one of the histidine genes. How would you go about isolating the wild-type allele of this gene?

6. A scientist finds an obscure bacterium with an enzyme that catalyzes the breakdown of a toxic byproduct of computer-chip manufacturing. It has real commercial potential, but it is made in very small quantities and the bacterium cannot be grown very easily. How should he begin to think of realizing the commercial potential of this discovery?

7. What attributes of a plasmid make it a suitable cloning vector?

8. A new restriction enzyme is discovered that recognizes an 8-base restriction sequence. About how many fragments of the *E. coli* genome would you expect if you digested it with this enzyme?

Answers

CHAPTER 1
ANSWERS TO REVIEW QUESTIONS

1. DNA is able to (i) specify replicas of itself; (ii) carry linear information of any complexity; and (iii) mutate.

2. There will be 1024 (2^{10}) cells.

3. On the 1' positions of the sugars, deoxyribose.

4. The nucleotide pair, of which there are four types: A-T, T-A, G-C, and C-G. (Note that, in terms of sequence information, there are four base pairs, although there are only two in terms of chemistry.)

5. Lagging-strand synthesis occurs because nucleotide chain extension is restricted to the addition of a deoxynucleotide 5' phosphate to the 3' OH group that defines the 3' end of the chain. Therefore, if the antiparallel strands at a replication fork are to be synthesized roughly in unison, they must be synthesized in opposite directions. Whereas one strand can be replicated toward the fork, the other, the lagging strand, must wait for enough single-stranded DNA to emerge from the fork and serve as a template for synthesis away from the fork.

6. The code for the ability to make the polysaccharide around the cell lies in one molecule, DNA, whereas the characteristic is the expression of this code, leading to the synthesis of a polysaccharide.

7. A primer is needed because DNA polymerization cannot occur without a 3' OH end of a polynucleotide, to which new deoxyribonucleotides are added. Either DNA or RNA can prime synthesis, but new strands begin with a short RNA chain.

8. It is present twice: once in each nucleotide chain, but in complementary forms.

9. Scientists thought proteins were so complex, especially in three-dimensional structure, that no simpler molecule could specify their structure. There was then no clear evidence that proteins were made up of linear, unbranched, one-dimensional sequences of amino acids. In fact, there were no alternative candidates for equally or more "complex" macromolecules. Most scientists changed their minds only (a) when DNA alone was shown to carry genetic information between generations or in transformation; and (b) when DNA structure was shown to be free to vary in nucleotide sequence and thereby able to carry information. By that time, enough was known about the structure of proteins to make

plausible conjectures about the translation of genetic information into protein structure. The final step was the demonstration that the three-dimensional structure of certain proteins could be achieved spontaneously by the protein, once it had been made as a linear chain of amino acids.

10. Avery realized that his purified DNA (a) could carry information (for the S character, missing in the R mutant); and (b) could duplicate after being incorporated into the transformed cell. That a molecule could carry mutable information and could duplicate defined it as genetic material—at least for the S character.

11. Hydrogen bonds are hydrogen atoms shared between specific atoms of the nitrogenous bases. There are two between A and T, and three between G and C. Hydrogen bonds are individually weak, but their large numbers in a DNA molecule impart stability to the duplex.

12. Both theta and sigma modes refer to replication of circular DNA molecules. In the theta mode, replication requires one or two RNA primers to form at the origin of replication. Neither nucleotide strand has to break at the origin of replication, and replication continues uni- or bidirectionally around the circle until two copies of the double-stranded circle are formed. Replication then terminates. In the sigma or rolling-circle mode of replication, one strand breaks and peels away from its complementary strand, thereby making a replication fork. Replication maintains one double-stranded circle from which a single- or double-stranded, linear DNA diverges. The latter becomes longer (often coming to have many tandem repeats of the information in the original circle) as the circle replicates at the fork.

13. Restriction enzymes recognize specific, short sequences in DNA and are used to cut DNA at these points. A pure preparation of a restriction fragment from the DNA of a complex organism can be made only by cloning methods in which restriction fragments are ligated ("inserted") into plasmids. The heterogeneous population of plasmids is introduced into a population of bacteria, and specific clones of bacteria carrying the plasmid of interest are selected by one of a variety of detection methods. The plasmids from an individual clone (all copies being identical) are isolated from the bacteria and cut with restriction enzymes; the fragment of interest that is liberated can be purified.

14. The probability that one of the four bases will occupy a given position in a DNA strand is 0.25. A specific sequence of four nucleotides will occur at an *average* frequency of $(0.25)^4$, or 0.00390625 of 4-base sequences. The reciprocal of this is 256 (1 in 256 bases). The use of fractions is an easier way to calculate this frequency: $1/4 \times 1/4 \times 1/4 \times 1/4 = 1/256$.

CHAPTER 2
ANSWERS TO REVIEW QUESTIONS

1. Synthesis of a DNA nucleotide chain needs not only a template, but a primer with a 3′ OH group to which new nucleotides can be added. DNA polymerase can therefore never copy the 3′ end of a template because there is nowhere to place the primer needed to copy it. Thus the ends of the new chains would become shorter each generation. In prokaryotes, there are no chromosome ends because the chromosomes are circular. In eukaryotes, the linear chromosomes have telomeres. They are made of special simple DNA sequences that the enzyme telomerase makes by itself and replaces as the telomeres become shorter by incomplete replication.

2. The centromere is a segment of DNA of the chromosome bound to proteins making up the kinetochore. These proteins in turn bind the spindle fibers which orient the chromosomes on the mitotic spindles and draw sister chromatids apart during division. In prokaryotes, a point of attachment of DNA to the inner surface of the cell membrane

becomes duplicated in cell division. The duplicate points, each attached to one of the daughter molecules, become separated on the membrane during growth of the cell, and the daughter DNA molecules move apart as they are replicated. A cell membrane and wall grow between the daughter molecules after they are complete, thereby separating them into the two daughter cells that form.

3. The nucleosome structure is an octamer of protein subunits having one pair each of the histones H2A, H2B, H3, and H4. DNA is wrapped around the globular octamer.

4. Euchromatin is less condensed and more likely to be active in forming gene products. Heterochromatin, in contrast, is inactive, condensed, and remains so after repeated cell divisions. Satellite DNA is repeated-sequence DNA that can be separated from the bulk of genomic DNA owing to its different base composition. Much of the satellite DNA is in the form of heterochromatin, especially near centromeres.

5. Chromosome replication (DNA synthesis) takes place in the S phase; mitosis takes place in the M phase. In the cell cycle, these two phases are separated by "gaps" in time: G_1, between M and S; and G_2 between S and M. Thus, the time at which a cell replicates its chromosomes is quite distinct from that at which it distributes them to daughter cells.

6. A centromere is the point on a chromosome at which spindle fibers attach during mitotic anaphase. Centrioles are the organizing centers of the spindle apparatus at the two poles of the cell during mitotic cell division.

7. A prokaryote lacks a nuclear membrane, has a single circular chromosome consisting of naked DNA, and has few or no membrane-bound organelles. A eukaryote has a nuclear membrane; multiple, linear chromosomes consisting of DNA, histones, and other proteins; and membrane-bound organelles such as mitochondria, chloroplasts, Golgi bodies, and vacuoles.

8. Bacterial chromosomes are circular, whereas those of eukaryotes are almost always linear and two-ended. Bacterial chromosomal DNA is "naked" (relatively few molecules are strongly bound to it), whereas eukaryotic chromosomal DNA is bound strongly to histones, which greatly condense the DNA. Both types of organism have nonhistone proteins, many of which control gene activity, loosely bound to their DNAs.

9. Haploids have one set of genes (and one set of chromosomes); diploids have two. In diploid organisms, one of the two sets comes from the male parent and the other from the female parent.

CHAPTER 3
ANSWERS TO REVIEW QUESTIONS

1. Haploids have a haploid vegetative phase in which cells have only one set of chromosomes. In diploids, the vegetative phase has two sets. Haploid organisms form gametes by simple differentiation of haploid cells. The diploid zygote formed by haploid organisms in fertilization soon undergoes meiosis. Diploid organisms form meiotic cells ("germ cells") by differentiation of diploid cells, and these undergo meiosis. Diploid animals form gametes directly from the meiotic products, while plants form a rudimentary haploid gametophyte after meiosis which produces gametes after a few mitotic divisions.

2. Haploid gametes fuse to form a diploid zygote. This is called fertilization. A diploid cell—either the zygote or mitotic derivatives—undergo meiosis to form four meiotic products. Gametes are derived from these meiotic products immediately in diploid organisms or after many cell divisions in most haploid organisms.

3. Any eukaryotic cell, haploid or diploid, can undergo mitosis, while meiosis takes place only in diploid cells. In mitosis no pairing of homologous chromosomes takes place, even in diploid cells, while chromosome pairing is one of the most characteristic features of meiosis. Mitosis consists of a single cell division preceded (in S phase) by DNA synthesis. Meiosis is a series of *two* cell divisions that follow a single round of DNA synthesis (premeiotic S phase). Mitosis maintains the ploidy of cells (either 1N or 2N); meiosis reduces the ploidy from 2N to 1N.

4. Homologous chromosome pairing is one of the most fundamental features of Prophase I, assuring that the diploid state is ultimately reduced to the haploid state by the end of meiosis. The exact pairing also assures the precision of recombination of homologous chromosomes by crossing over. Note: DNA synthesis takes place *before* Prophase I.

5. A *bivalent* consists of two paired homologs. The term *homolog* refers to one of two chromosomes carrying the same genes in a diploid cell, one derived from the maternal and the other from the paternal parent. A *chromosome* is one or two linear molecules of DNA, including a centromere, and associated with proteins. It may have a single copy of the DNA (in G_1) or it may have two copies (in G_2). In the latter case, there are two *chromatids* joined at the centromere, and these are visible in prophase of mitosis and meiosis.

6. Anaphase I, if no crossing over has occurred between the gene in question and its centromere. If crossing over occurs, segregation is delayed to the second anaphase: a phenomenon called "second-division segregation" (*see* Chapter 9).

7. A *gene* is a segment of DNA encoding a type of protein (or, in general terms, a character) of an organism. The word is more general than *allele,* since alleles are specific alternative forms of a gene. Alleles are related in three ways: they are related by mutation (and can be traced back to a common ancestor); they lie at the same place (locus) on homologous chromosomes; for that reason, they always segregate from one another when heterozygotes carrying different alleles undergo meiosis.

8. This problem forces you to recognize the ability or inability to grow without adenine as alternative characters that segregate in a 1:1 ratio. These meet the criteria of allelism: ade^+ and ade^- are alleles of the *ade* gene. There is only one gene in the cross, and the ade^+ allele is dominant; the diploid formed in the cross has the phenotype of the ade^+ parent. One of the two bivalents in Figure 3–2 represents the way a bivalent can be depicted at Metaphase I.

    Note: In this and most other problems involving yeast to follow, we will present results that assume that a random collection of meiotic products (ascospores) is analyzed. While this is occasionally done, most yeast crosses are analyzed by dissecting a number of individual tetrads. This has the advantage of distinguishing ascospores from diploid, parental cells, and of yielding more genetic information than random spores. For our purposes, the more general and elementary principles are best illustrated with random spores, the aggregate of many meiotic events.

CHAPTER 4
ANSWERS TO REVIEW QUESTIONS AND PROBLEMS

1. The appearance of a new phenotype shows that ebony is recessive and that both parents are heterozygous for it. The ratio of progeny (3:1) indicates that only one gene underlies the different phenotypes.

2. The F_2 genotypic ratio is 1 BB : 2 Bb : 1 bb. Heterozygotes are 2/3 of the black F_2.

3.   The 1:1 ratio in the F_1 shows segregation is occurring in only one parent, the short-tailed mouse. (The other parent ["true-breeding"] is known to be homozygous.) Because the short-tailed mouse is heterozygous, the short-tail character is dominant or semidominant. Thus if the first mating were $Tt \times tt$ (T being the short-tail determinant), the results are understandable. The second mating would then be $Tt \times Tt$. We can account for the tt (normal) and the Tt (short-tailed). The TT genotype is evidently lethal, leaving only the 2:1 ratio of the remaining classes. With respect to viability, the T allele is semidominant, since the heterozygote is not as severely affected as the TT homozygote.

4.   Testcross the male to a retinal atrophic female. The appearance of mutant pups would show the stud was a carrier. If none appear in several litters, it is unlikely that the stud is a carrier.

5.   One half. This problem should force you to translate the statement into symbolic form: $wr^+/wr \times wr/wr$.

6.   In diploids, one cannot determine phenotypes of meiotic products, because they do not express characteristics of the vegetative, diploid stage. Therefore, segregation of alleles is seen only in the following diploid phase. This may be an F_2 (3:1 ratio of phenotypes for a dominant-recessive pair of alleles) or a testcross progeny of a monohybrid (1:1 ratio, reflecting the two types of gametes produced by the hybrid).

CHAPTER 5
ANSWERS TO REVIEW QUESTIONS AND PROBLEMS

1.   Equal numbers of four genotypes: *AaBb, Aabb, aaBb* and *aabb*.

2. $1\ AB : 1\ Ab : 1\ aB : 1\ ab$.

3.   (i) $1/2 \times 1/2 \times 1/2 = 1/8$; (ii) $3/4 \times 3/4 \times 1/4 = 9/64$; (iii) $1/4 \times 1/4 \times 1/4 = 1/64$; (iv) $1/4 \times 1/2 \times 1/4 = 1/32$

4.   The tan and red phenotypes are alternative traits because they emerge in a 1:1 ratio among the progeny. The diploid is tan, showing that this is the dominant character. The cross would be diagrammed by using red^+ as a symbol for the wild type allele determining the tan character (in practice, yeast geneticists use capital letters to signify the dominant allele; hence *RED*) and the tan allele would be red^- (in practice, *red*). The 1:1:1:1 ratio of the four possible types of progeny show that mating type and color segregate (assort) independently, and that at least *two* types of Metaphase I arrangement and two types of tetrads must be included in the diagram (*see* Fig. 5–1).

5.   The smooth, yellow F_2 plant that gave rise to smooth, yellow and smooth, green progeny was clearly *not* segregating for wrinkled, and therefore was homozygous for the smooth allele. Because the alternative yellow and green progeny appeared in a 3:1 ratio, the F_2 plant was heterozygous for smooth and green alleles. The genotype can be represented *AABb*.

6.   The basis of independent assortment is that any two bivalents line up independently of one another at Metaphase I of meiosis. For genes on different bivalents, an allele of one gene may segregate with either allele of the other gene, and with equal probability. Therefore, all four combinations of alleles, two parental and two recombinant, will be equally frequent among a population of meiotic cells.

7.   (i) $1/2 \times 1/4 \times 1/2 = 1/16$; (ii) $1/2 \times 3/4 \times 1/2 = 3/16$; (iii) $1/2 \times 1/4 \times 1/2 = 1/16$; (iv) $1/2 \times 3/4 \times 1/2 = 3/16$. (For each gene, the proportions of progeny can be predicted from Table 4–1, Chapter 4.)

8. The expected values for a progeny of 48, would be 27, 9, 9, and 3 for the categories as given in the problem. A χ^2 test must exclude the last value, because the expected number is less than 5. Using only the first three categories, we obtain a χ^2 value of 1.44. With 2 degrees of freedom (only 3 categories are being considered), $P = 0.5$–0.6.

9. In this haploid organism, independent assortment should yield the four phenotypes in a 1:1:1:1 ratio. The χ^2 value is 9.983; with 3 degrees of freedom, $P = 0.02$—unacceptably low. The segregations of red^- and red^+ (1417:1335) approximates a 1:1 ratio with an acceptable P value, but the segregation of gal^- and gal^+ (1441:1311) does not. Among the two categories of gal progeny, the segregation of red is 1:1. The data as a whole indicate that the genes assort independently, but that the gal^- progeny are differentially lost from the progeny; they appear to be less viable.

10. It is simplest to calculate first the frequency of the category without boys: those with four girls. If the probability of having four girls is $(0.5)^4 = 0.0625$, then the remainder have a combined frequency of 0.9375. This frequency can also be calculated more laboriously by expanding the binomial $(p + q)^4 = p^4 + 4p^3q + 6p^2q^2 + 4pq^3 + q^4$ and noting that 15/16 (0.9375) of the distribution has one boy.

11. One third, since the frequencies of MFF, FMF and FFM families are equal.

12. Of four-child families, 6/16 have two boys and two girls; only 1/6 of such families will have the birth order MFFM. Therefore, 1/16 will have that particular birth order. The same answer can be derived as 0.5^4.

CHAPTER 6
ANSWERS TO REVIEW QUESTIONS AND PROBLEMS

1. 32 (4 × 4 × 2). If one codon is assigned as a termination codon, only 31 amino acids can be encoded with this code.

2. The **primary** structure of a protein is the unbranched sequence of amino acids of the polypeptide(s) of which it is made. The **secondary** structure forms by natural interactions of the peptide backbone(s), forming α-helices and β-sheets. The **tertiary** structure is determined by the interactions of the R groups, often between amino acids distant from one another in the polypeptide sequence, and it is both constrained by and constrains the secondary structure. The **quaternary** structure refers to the number and types of folded polypeptide chains in the mature protein: the quaternary structure may be monomeric (consisting of a single polypeptide), homomultimeric (aggregates of the same type of polypeptide), or heteromultimeric (aggregates of polypeptides encoded by different genes). In all cases, the quaternary structure is determined by the chemical interactions of R groups at the interfaces of the folded polypeptides.

3. Enzymes are, as a rule, proteins (although some RNAs act as enzymes). Enzymes catalyze biochemical reactions by binding the molecules that they will transform (the substrates) and facilitating chemical changes favored by the energy content of the substrate(s). The products are then released. If energy is required in a reaction, other substrates must provide it, or the product must be consumed rapidly so that a reverse reaction does not oppose the forward reaction. A hallmark of enzymatic reactions is their specificity: an enzyme catalyzes only one or a small number of reactions, because it recognizes the shape of only a few substrates. Another type of protein is structural, rather than catalytic, exemplified by collagen and hair (external proteins) and by cytoskeletal proteins such as actin, tubulin, and histones.

4. Synonymy in the code refers to the fact that more than one codon can specify a single amino acid. While this does not impart ambiguity as nucleic acids are translated, the

codons used to encode a protein of known amino acid sequence cannot be deduced without ambiguity.

5. 160,000 (20^4).

6. 216 (6^3).

7. RNA has only one, not two, nucleotide chains; its sugar is ribose, rather than deoxyribose; and the nitrogenous base uracil is found in place of thymine.

8. Nine: there are three alternate bases for each of the three bases of this codon; $3 + 3 + 3 = 9$.

9. Five: Met (AUG), Trp (UGG), Val (GUG), Ser (UCG), and Phe (UUC, UUU). Codons CUG and UUA code for Leu and would not yield a substitution, and UAG, the last of the nine possible alternatives, is a stop codon.

10. 5′ AGGCUUACUGGCCU 3′.

11. Both nonsense mutations (mutation of a sense to a non-sense codon) and frameshift mutations (insertion or deletion of a single base pair) can abolish gene function. The first leads to termination of reading; the second shifts the reading out of the proper reading frame in a coding sequence.

12. (i) Many base-substitutions lead to changes to a codon synonymous with the original codon. This is particularly the case if they happen in the third position of codons, as inspection of the code table (Fig. 6–4) will reveal. (ii) A change from a sense codon to a nonsense (chain-terminating) codon, if it is not at a position encoding the very C-terminal end of the protein, will lead to a truncated protein—in most cases lacking function and stability.

13. The NN child: MM and MN parents cannot yield such a genotype.

14. MM and MN; the woman (genotype *MN*) produces *M* and *N* gametes, the man (genotype *MM*) only *M* gametes.

15. A dominant allele imparts the same phenotype to the dominant homozygote and the heterozygote. This is easily observed in matings that yield a hybrid from true-breeding parents differing in one pair of alleles. Homozygotes for a semidominant allele, however, do not have the same phenotype as the heterozygote; the latter is intermediate (if not exactly so) between the phenotypes of the two homozygotes.

16. Laboratory mutations (and in fact many natural mutations) are recessive because they simply inactivate a gene, and the activity of the wild-type allele is sufficient in single dose to impart a normal phenotype to heterozygotes.

CHAPTER 7
ANSWERS TO REVIEW QUESTIONS AND PROBLEMS

1. Two complementary recessive genes are involved. One parent is *AAbb,* the other is *aaBB*. Both mutations *a* and *b* are recessive, but affect the same pathway of pigment formation. The F_1 progeny (genotype *AaBb*) are all capable of making purple pigment. However, only 9/16 of the F_2 will carry at least one dominant allele of both *A* and *B* genes. The other 7/16 will be unable to make pigment because one or both of the genes in question is homozygous for the recessive allele. You may have been tempted to think that the F_2 ratio was 1:1, even though the problem statement makes this hard to rationalize. A χ^2 test of the data

with this ratio yields a χ^2 value of 14.06 and a P value, with 1 degree of freedom, less than 0.001. A test of the 9:7 ratio, however, yields a χ^2 value of 1.59 and a P value of 0.2–0.3.

2. This problem involves duplicate genes: both must be mutant in order to impair color. Thus the parents are *aabb* (white) and *AABB* (red), with *A* and *B* dominant. The F_1 (*AaBb*) is red, and only the 1/16th *aabb* class of the F_2 is white, because only this class lacks dominant alleles of both genes. A test of the 15:1 hypothesis yields a χ^2 value of 2.67 and a P value, with one degree of freedom, of 0.1–0.15.

3. The F_2 ratio suggests that the F_1 was heterozygous for two genes with duplicate functions. The parents were probably (and you must infer this) homozygous for the two dominants in one case and the two recessives in the other. Thus a testcross of the F_1 would have the form *AaBb* × *aabb*; among the four progeny genotypes, only one (*aabb*) would be white. The other three, *AaBb, Aabb,* and *aaBb,* would have at least one gene able to impart color. This problem is deceptive in giving you a 3:1 ratio—but in a testcross, in which it has a different interpretation than in an F_2.

4. Red and white are semi-dominant, since the F_1 is pink. The segregation of the colors confirms this; it is 1:2:1 in the F_2. The abnormal flower shape is recessive, since the F_1 does not show it, and the expected 3 normal : 1 abnormal is seen in the F_2. Clearly, the genes are independently assorting, because the 1:2:1 ratio of colors appears in both classes of flower shape. (This is a vital "checklist" trick: take each gene or character individually before you try to solve the problem as a whole.)

5. 3/16 short yellow; 6/16 short cream; 3/16 short white; 1/16 long yellow; 2/16 long cream; 1/16 long white. Note that this is a 3/4 + 1/4 distribution for short/long multiplied by a 1/4 + 1/2 + 1/4 for white/cream/yellow.

6. You are asked to predict the frequency of the double heterozygote among the F_2 progeny, the cross being *AaBb* × *AaBb*. The answer is 1/4, and can be derived by recognizing that *Aa* × *Aa* produces 1/2 *Aa;* similarly, 1/2 the progeny will be *Bb*. Thus 1/2 × 1/2 = 1/4 will be medium pink (*AaBb*).

7. The fungus is haploid, and the progeny reveal a 1:1 ratio of conidial to aconidial. Moreover, the conidial progeny reveal a 1:1 ratio of colored to albino. Thinking realistically, you will understand that progeny unable to form conidia cannot reveal the state of the gene controlling conidial color; in other words, the aconidial trait is epistatic to the color trait. A χ^2 test would compare the results with an expected 1 normal : 1 albino : 2 aconidial ratio. The χ^2 value is 0.776; $P = 0.7$–0.8, with two degrees of freedom.

8. This problem must be approached with some imagination. Upon finding that the F_1 plants are purple, one may ask whether this is the action of two semidominant alleles of one locus (red + blue = purple), or whether two different genes contribute the two colors to the F_1 phenotype. The F_2 progeny answers this question: if one locus were involved, the proportions would have been 1 red : 2 purple : 1 blue. The 9:3:3:1 ratio of purple, red, blue and white, respectively, suggests that two genes are involved; that red and blue are homozygous for *one* of each of two recessive genes; and that white flowered plants are double-recessives. This hypothesis is best formulated with symbols: *A_B_* would be purple; *A_bb* would be red; *aaB_* would be blue; and *aabb* would be white. The original cross would therefore have to be *AAbb* × *aaBB*. The final question was why true-breeding red × white yielded no purple plants. If you see the cross as *AAbb* × *aabb,* the answer is simple: there is no dominant *B* allele in either parent.

9. (This problem is related to the previous problem, but should be solved independently.) The initial cross tells you that the white color is recessive in some sense: it does not appear in the F_1. This cross does not tell you how many genes there are; there could be just one, with alleles determining purple and white. However, the backcross appears to be a testcross (remember that the white parent is true-breeding, and is apparently recessive). In equal proportions, four colors appear, a signal that two genes each with two alleles might be assorting independently. What could these genes be? Some attention to the actual phenotypes is

needed: purple is a mixture of red and blue; we see a 1:1 segregation of progeny with blue pigment (blue and purple) and without blue pigment (red and white), and a 1:1 segregation of progeny with red pigment (red and purple) and without red pigment (blue and white). Given this, we should expect a 9 purple : 3 red : 3 blue : 1 white ratio in the F_2 progeny.

10. By identifying alternative characters (for height and for color), you notice in the progeny that tall, medium and short segregate 1:2:1, respectively, as do red, purple and blue. The ratio is characteristic of genes with semidominant alleles. Notice also that the two genes assort independently: the 1:2:1 ratio for one gene prevails in each phenotypic category determined by the other gene. This allows you to formulate the genotype in symbols: the first cross was $AABB \times aabb$ (realize that the alleles are semidominant) to yield $AaBb$. If this F_1 is back-crossed to the short red parent ($AABB$), the cross would yield equal numbers of short red ($AABB$), short purple ($AABb$), medium red ($AaBB$), and medium purple ($AaBb$) progeny.

11. Dominance is a relationship between alleles of the same gene, in which the dominant allele obscures the presence of the recessive. Epistasis is a relation between a mutation in one gene that obscures or modifies the expression of differences between alleles at another gene.

12. In this case, an apparently simple cross yielded a large majority of mutant progeny. This can only mean that the normal wild-type strain (not requiring arginine) is a rare genotype among those produced in the cross, and that the arginine-requiring strain harbored more than one "arg^-" mutation, any one of which could impose an arginine requirement. If the parental arginine-requiring strain carried two mutations, only one-fourth of the progeny would be wild-type. If it carried three arg^- mutations, we would expect one-eighth of the progeny be wild-type. A χ^2 test of a 1:7 ratio gives a χ^2 value of 1.96 and a P value of 0.15–0.20 with one degree of freedom.

13. The two parents were homozygous for two different recessive genes, each of which caused deafness in the homozygous state. This is a simple case of complementary genes.

CHAPTER 8
ANSWERS TO REVIEW QUESTIONS AND PROBLEMS

1. One-half. Color blindness is sex-linked; the woman must be heterozygous for the mutation, and therefore one-half her sons will receive the mutation. It will be expressed, owing to the hemizygosity of males. The mating would have the same outcome even if color blindness were autosomal. Do you see why?

2. In the F_1, all progeny will have normal body color. The females will have red eyes, the males will have white eyes. In the F_2, ebony will segregate 3:1 in each sex, while white will segregate 1:1 in each sex. The ratio among the progeny will thus be 3/8 red-eyed, normal-bodied; 3/8 white-eyed, normal-bodied; 1/8 red-eyed, ebony-bodied; and 1/8 white-eyed, ebony-bodied.

   Note: The statement of the problem does not tell you that the ebony males are normal in eye color, nor that the white-eyed females have normal body color; only mutant characters are stated. Although this problem is not particularly prone to misinterpretation, you must always be conscious of the need to recognize explicitly the allelic states of all relevent genes in both parents.

3. 2 females : 1 male

4. Female. The mother, being mahogany, must be AA. Her calf must have received an A allele from her, but being red, must have received the a allele from its father. If it was heterozygous and red, it must be female.

5. Both traits are sex-linked, the mother being heterozygous for both traits. The father is normal. The kinds of sons indicates that the mutations are on different homologs of the mother's sex chromosomes (the X's). The couple could have a normal son only if the mother produced an egg (meiotic product) carrying a recombinant X chromosome, with a wild-type allele for each gene (*see* Chapter 9). The parents are:

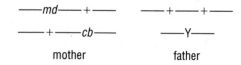

mother   father

6. Note that in this problem, *w* is epistatic to *st*. (i) In the first mating, the F_1 males and females will both be red-eyed, and the F_2s will be as follows: Females: 3/4 wild type and 1/4 scarlet; Males: 1/2 white, 3/8 wild type, 1/8 scarlet. (ii) In the second mating, the F_1 will yield white eyed males and red-eyed females. The F_2s of both sexes will be as follows: 1/2 white, 3/8 red, and 1/8 scarlet.

7. Large-comb males and females without combs. Remember that the heterogametic sex in fowl is the female.

8. The ebony males of the F_2 are *eb/eb ct$^+$/←*. Taking one gene at a time: the ebony character is absent from the F_1 and present in 1/4 of the F_2 of both sexes. This identifies it as an autosomal gene. Independently, the cut-wing character is distributed differently in the sexes of the F_2. This suggests that it is sex-linked, and the crosses bear this out: it appears in a 1:1 ratio with normal in the F_2 males.

    Once again, notice that only specific mutant characters are listed. You need to infer the normal characters as alternative traits for the mutants, then assign symbols, and then relate them to patterns of inheritance.

9. Color-blind sons and normal daughters.

10. The appearance of ebony and normal body in a 1:1 ratio among both males and females shows that one parent is segregating for this character and the other is not; it is a testcross ratio of an autosomal gene. (You must know that ebony is recessive to decide that the normal-bodied female is heterozygous.) The appearance of white eyes in males, but not in females, and in a ratio of 1:1 with the red-eye alternative, suggests that it is sex linked, and that the female is heterozygous for it. The male must be hemizygous for w^+, since he has normal eyes. Thus the cross is: $eb^+/eb \ w^+/w \times eb/eb \ w^+/-$.

11. The woman initially should not have worried about her daughter; the normal vision of her husband assures daughters of normal color vision. The daughter, however, is a carrier (heterozygote), and the mother should tell her daughter that her sons have a one-half probability of being color-blind. The daughter's normal husband will render all daughters normal, though half will be carriers.

12. If Barred females (*B*/←: the female is heterogametic in fowl) are mated to normal males (*b/b*), all female progeny will be normal (*b*/←) and all males will be Barred (*B/b*).

CHAPTER 9
ANSWERS TO REVIEW QUESTIONS

1. (i) Anaphase I or II; (ii) Pachytene; (iii) Metaphase I; (iv) Zygotene; (v) Anaphase II.

2. A map unit is equal to 1% recombination, as measured among the haploid products of meiosis.

3. In *Neurospora,* all products of the first division are in one or the other half of the ascus. In second-division segregation for a given gene, each half of the ascus contains both alleles of the gene. This could happen only if the exchange between homologs took place when there were four copies of the chromosome, of which two recombined, and two did not.

4. NPD tetrads segregating for unlinked genes appear as the outcome of one of the two possible bivalent orientations during Metaphase I, there being two bivalents involved. NPD tetrads for genes on the same chromosome (and therefore on the same bivalent) require that all strands be recombinant. This requires two crossovers in a 4-strand, double crossover tetrad. See Figure 9–3.

5. The association of crossing-over with conversion suggested that they originated in the same event. In fact, common to both is the formation of one or two Holliday structures. The repair of heteroduplex DNA following the establishment of a Holliday structure is the origin of gene conversion, and the resolution of the Holliday structure(s) after isomerization is the origin of crossing-over. The isomerization of the Holliday structure almost randomizes the flanking markers between parental and recombinant arrangements, thereby yielding about 50% recombination among tetrads in which gene conversion occurs.

6. (i) In haploids, linkage tests are made by mating parents that differ in two or more genes and determining the frequency of parentals and recombinants among the progeny. (ii) In diploids, parents differing in two or more genes must be mated to make a hybrid. The hybrid is then testcrossed in order to estimate the frequencies of parental and recombinant meiotic products.

CHAPTER 9
ANSWERS TO PROBLEMS

1. $A — 8 — D — 12 — B$ (The map $B — 12 — D — 8 — A$ is equally valid.)

2. This is a testcross, so the ratio of meiotic products will be revealed in the phenotypes of the progeny. The progeny will be 86% parental (43% wild-type, 43% scarlet, spineless) and 14% recombinant (7% scarlet, 7% spineless).

3. *ABD, Abd, aBd,* and *abD.*

4. $A — 2.8 — D — 4.4 — B$. The coefficient of coincidence is about 0.9. (Note that the total of progeny is 900, not 1000.) The value of the coefficient of coincidence is not secure, owing to the appearance of just one double-crossover product. A larger progeny would allow us to measure the coefficient of coincidence with greater accuracy.

5. Ten (10), calculated as the following product: $0.1 \times 0.2 \times 0.5 \times 1000$.

6. The class asked for is one-half of one of the single-crossover classes. The number of *A-B* recombinants, including doubles, will be 10%, or 100 progeny. Of these, 20 will be double crossovers, leaving 80 in the single-crossover category *ABd + abD.* Of these 80, one-half, or 40, will be *ABd.*

7. $pr — 22.2 — v — 43.3 — bm.$ c.c. $= 0.8$

8. Start by recognizing that the phenotypic distribution of both sexes will be the same. Then calculate double crossovers: $0.077 \times 0.29 \times 1000 = 22$ (*ct + f* and + lz +). The total crossovers in region I (*ct-lz*) will be 77, of which 22 are doubles, leaving 55 single crossovers (*ct + +* and *+ lz f*). Similarly, the total crossovers in region II (*lz-f*) will be 290, and the single crossovers will be 290 − 22, or 268 (*ct lz +* and *+ + f*). The remainder of the progeny (655) will be equally divided between the parental classes, *ct lz f* and *+ + +.*

9.   First, arrange genotypes with reciprocal pairs together and sum their numbers.

$$AbdE + aBDe = 903 \qquad Abde + aBDE = 47$$
$$abDe + ABdE = 902 \qquad AbDe + aBdE = 48$$
$$ABde + abDE = 48 \qquad abde + ABDE = 3$$
$$ABDe + abdE = 47 \qquad AbDE + aBde = 2$$

Second, notice that the "parentals" (those with more than 900 progeny) are four classes differing only in the distribution of the B and b alleles; otherwise all of them are either AdE or aDe. The equal frequency of B and b alleles in these classes shows that B assorts independently of A, D, and E. This is borne out in all other frequency classes. The groups must be re-summed without regard to B/b: $AdE + aDe = 1805$; $Ade + aDE = 95$; $ADe + adE = 95$; and $ade + ADE = 5$. This is a simple three-point cross with D/d in the middle, flanked by A and E 5 map units on either side:

$A — 5 — D — 5 — E;$ coefficient of coincidence is 1.0. B lies on another chromosome.

10.   The double-recessive phenotype represents a single genotype, $rrdd$. The original cross tells you that the dihybrid would be Rd/rD, so that the F_1's rd gametes would have to form by crossing over. In a self-cross of the dihybrid, the genotype $rrdd$ would arise only by fertilization of a rd ovule and a rd pollen grain. You are told that the frequency of $rrdd$ is 0.01, a frequency that is the product of separate (and in this case necessarily equal) probabilities. Therefore the rd gametes have a frequency of 0.1 and rd/rd zygotes will appear in a frequency of 0.01 (0.1×0.1). The rd gametes are one-half the recombinants of an Rd/rD plant; to these must be added an estimated 0.1 RD gametes to get the map distance, $0.2 \times 100\%$, or 20 map units. This example shows why an F_2 is a poor progeny with which to determine map distance.

11.   The genotype and gene order of the trihybrid are AdB/aDb, using the parental (ABd and abD) and double crossover (abd and the nonexistent ABD) categories. The map distance between A and D is $(12 + 12 + 1)/1000 \times 100\%$, or 2.5 map units; the distance between D and B is $(19 + 20 + 1)/1000$, or 4.0 map units. The map is thus $A — 2.5 — D — 4.0 — B$. The coefficient of coincidence is 1.0, dividing the observed 0.001 double crossover progeny by the product (0.04×0.025). There is thus no interference, although the number of double crossovers on which the conclusion is based is too low for confidence.

12.   The data should be analyzed by recognizing the equality of the eight possible classes. This can only mean that the three genes assort independently, and any meiotic product will be 1/8 the progeny. The fraction is understood by knowing that each pair of alleles distributes itself in a 1:1 fashion. Thus the fraction of each genotype among meiotic products involving three unlinked genes is $1/2 \times 1/2 \times 1/2$, or 1/8.

13.   This problem is worked out here fully as the reverse of a standard mapping problem, using shorthand phenotypic designations, and the answers to the specific problem pointed out. First, if there is no interference, the number of double crossover progeny can be predicted simply from the map distances given (10% and 20%). When converted to frequencies and multiplied, we have the expected number of double crossovers, 20 of 1000 progeny. Of these 20 double crossovers, 10 are RdY (answer to the first question) and 10 are rDy. The 1000 progeny should have 10% progeny (or 100 in all) that have crossovers between R and D. 20 of these are accounted for among the double crossovers, leaving 80 single crossovers, 40 Rdy and 40 rDY. By similar means, we find there are 180 single crossovers in the second region, 90 rdY and 90 RDy. The remainder of the 1000 progeny will be parentals, RDY and rdy.

If the coefficient of coincidence were 0.3, there would have been 6 double crossovers ($0.3 \times 0.1 \times 0.2 \times 1000$) instead of 20. Of the 6, 3 will be RdY (answer to the second question) and 3 rDy. The number of single crossovers would have been 94 and 194 for the first and second regions, respectively, if the coefficient of coincidence were 0.3.

14. This problem asks how many *rrdd* zygotes would be expected in random fertilizations in which the frequency of *rd* gametes (in males and in females) is 0.1. This gametic frequency was determined in the testcross of the *Rd/rD* parent. The expected frequency of zygotes of the *rrdd* type would be 0.1×0.1, or 0.01 of the F_2 progeny. (This is the reverse of Problem 10, above.)

15. Take two genes at a time. Notice that *ab* and *AB* are more frequent than *Ab* and *aB* (160 vs. 38). *A* and *B* are linked (map distance = 19 map units). Is *C* linked to *A* or *B?* The answer is no; it assorts independently of both, as you will find when the *A-C* and the *B-C* combinations are totalled.

16. The first mating would give females heterozygous for both characters and males hemizygous for both mutant characters. Both, at this point, look as if they are sex-linked, owing to the fact that both characters are distributed differently in the two sexes; both white eye and cut wing are recessive, because the female is normal. The second mating confirms that the two characters are on one chromosome (the X) and, by reference to the genotypes of the F_1 flies, the similar distributions of the F_2 males and females are understood. (In effect, both chromosomes contributed by the F_1 male to the second mating are "testcross" chromosomes, unable to obscure the information contributed to zygotes by the F_1 female eggs.) The map distance between the mutations is $24/400 \times 100\% = 6$ map units.

17. The *A_bbdd* phenotype will be one of the single crossovers. One must first figure out the number of double crossovers, then the number of singles, then the number of the particular single-crossover genotype asked for. **Step 1.** Double crossovers that actually appear are $0.08 \times 0.20 \times 0.4 = 0.0064$, determined by multiplying the separate map distances (as probabilities) together, then multiplying the product by the coefficient of coincidence. **Step 2.** The *A_bbdd* phenotype is one of the two equally frequent crossovers in region I (between *A* and *B*); these will appear in the progeny in numbers that are not already accounted for by double crossovers. Total crossovers in region I are 0.08. **Step 3.** Using the numbers derived, single crossovers will be $0.08 - 0.0064 = 0.0736$; of these, one-half, or 0.0368, will be *A_bbdd*. Thus in a progeny of 1000, there will be 37 "*Abd*" phenotypes.

18. It is 0.5 map unit from its centromere: of the 1% tetrads that show second-division segregation, only 1/2 of the strands in them underwent recombination. Another way of looking at this is that map distance is the percent recombinant meiotic products, not the number of tetrads showing recombination. There are 400 meiotic products in every 100 tetrads; in the 1 tetrad (1%) showing second-division segregation, only 2 of the 4 chromatids are recombinant in this problem. As a percentage of meiotic products, this is $2/400 = 0.5\%$ recombinants.

19. The map is A — 0.5 — D — 5.3 — B. The coefficient of coincidence was not determinable because too few progeny were obtained (the expected double crossover progeny were $0.005 \times 0.053 = 0.000265$, or less than 1 out of 1000 progeny). In this problem, the double crossover genotypes had to be derived by elimination so that they could be used to establish gene order.

20. The map is *arn* — 6 — *tup* — 13 — *spd,* and is based on the *arn*[+] progeny. Three *arn*[+] classes are present in frequencies suggestive of parental (*spd1 tup4*) and two single crossover classes (*tup4* and wild-type). One of the *arn*[+] classes, *spd1*, is missing (it should be viable), suggesting that it is the double crossover category. We must write out the full genotypes of the presumed parental and double-crossover classes the get the map order, and the map distances are straightforward: $13/100 \times 100\%$ and $6/100 \times 100\%$.

The map distances could have been derived from all progeny, with the provision that 6 triple mutants (a number equal to the wild-type progeny, the complementary class) be added to one of the single crossover categories for the calculation. The values, calculated in this way, would be 6.19 and 11.9 units.

21. 50% PDs and 50% NPDs. Tetratypes, which require crossing over between one of the genes and the centromere, cannot arise in this cross.

CHAPTER 10
ANSWERS TO REVIEW QUESTIONS AND PROBLEMS

1–3. The following diagrams give the configurations asked for:

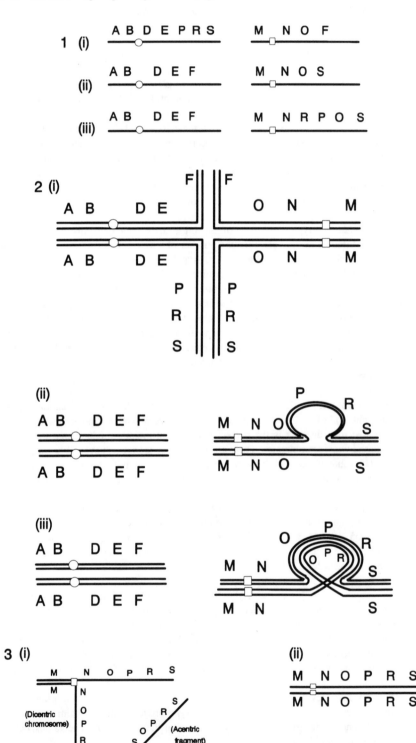

4.   The number of possible trisomic conditions (assuming only one chromosome is trisomic in each case) is equal to the haploid chromosome number. In humans, this is 23; in *Drosophila*, it is 4.

5.   Triploids maintain the same ratio of all genes (disregarding here male triploids) as is found in the diploid state; gene balance is not disturbed. Trisomics, however, have an improper ratio of the genes of one chromosome to those on all other chromosomes, and this may often be fatal.

6.   In normal people, only one sex chromosome (the X) is fully active; the Y is largely (though not completely) inactive, as is one of the two Xs of females. All additional X chromosomes are inactivated during early embryogenesis. Therefore, while the effects of extra sex chromosomes is clear in many cases, the effects are much milder than any autosomal trisomy.

7.   The types and proportions of gametes will be 2 *Aa* : 1 *aa* : 1 *A* : 2 *a*. Segregation at meiosis will be such that two chromosomes of the three go to one pole and 1 to the other. There are three arrangements of this kind (defined by which of the three homologs behaves as a univalent).

8.   That in somatic cells of many animals, all X chromosomes but one are inactivated. This applies to sex chromosome constitutions of XX (normal female), XXX, XXY, etc.

9.   (i) male; (ii) lethal; (iii) female; (iv) female; (v) female.

10.   By nondisjunction of Chromosome 21 in one of the early cleavage divisions of a fertilized egg, leading to a mixture of trisomic, diploid (and initially, monosomic) cells.

11.   A highly condensed, transcriptionally inactive X chromosome found in nuclei that have another, active X chromosome. In cells with more than two X chromosomes, all in excess of one become Barr bodies. In interphase, a Barr body lies on the inside of the nuclear membrane. It replicates late in S phase, and a copy is distributed in mitosis to each of 2 daughter cells, where it recondenses.

12.   (i) XX, X, XY, and Y. (ii) The zygotes (and their ratio) produced by fertilizations of the four gametes in (i) by X- or Y-bearing sperm would be: 1 XXX (lethal) : 1 YY (lethal) : 2 XX : 2 XYY : 3 XXY : 3 XY. In *Drosophila*, this would lead to equal numbers of females (XX, XXY) and males (XY and XYY).

13.   The nondisjunction yielding a zygote with two Y chromosomes could only come from a YY sperm. Such sperm can arise only at the *second* meiotic division. Nondisjunction at the first division will yield disomic sperm, but they will be XY not YY.

CHAPTER 11
ANSWERS TO PROBLEMS

1.   The data indicate that the wing characters are alternatives controlled by alleles of a sex-linked gene. If so, the parents are, with respect to this character, +/+ females × *c*/⟵ males. This would yield the pattern of inheritance shown, in which the cut character (recessive) fails to appear in the F_1 and appears only in the males of the F_2, where it segregates 1:1.

The white character also appears to be sex-linked, once you exclude the idea that it is not an alternative to brown. (This alternative—that brown and white are alleles—is not consistent with the appearance of wild-type F_1 females; they would probably be white or brown or an intermediate color.) Assuming white and its normal allele ("non-white") are sex-linked, the original parents are *w*/*w* females and +/⟵ males. Such parents would yield an F_2 in which both sexes segregated 1 non-white : 1 white.

The conclusion that the white and cut genes are both sex-linked means they are linked, in repulsion in the F_1 females and will not recombine freely. The map distance between them can be calculated if you correctly identify parentals and recombinants among the appropriate F_2 flies, the males. The males consist of 86 white and 86 cut (the parentals) and 14 white, cut, and 14 normal (the 28 recombinants) with respect to these characters. The linkage distance is $28/200 \times 100\% = 14$ map units. This is easier to visualize below, where genotypes are shown.

With respect to brown and its normal counterpart (red, the normal *Drosophila* eye color), you will see that non-white flies of both sexes in the F_2 segregate 3 red : 1 brown, as any autosomal gene normally does. However, only half the flies have colored eyes: white is epistatic to brown. We know enough to see that the original parents were bw^+/bw^+ females and bw/bw males.

Having reached this point, little is left but to predict carefully the genotypes of the F_1 and the F_2 flies and determine their phenotypes, recognizing that white is epistatic to brown, and cut is linked to white.

Parents:

> $w\,ct^+/w\,ct^+$ bw^+/bw^+ females and $w^+\,ct/\!\leftarrow$ bw/bw males.

F_1 flies:

> $w\,ct^+/w^+\,ct$ bw^+/bw females and $w\,ct^+/\!\leftarrow$ bw^+/bw males.

F_2 females:

> 98 white: $w\,ct^+/w\,ct^+$ and $w\,ct/w\,ct^+$, (each category segregates *genotypically* 1 bw^+/bw^+ : 2 bw^+/bw : 1 bw/bw).

> 102 colored eyes, genotypes $w^+\,ct^+/w\,ct^+$ and $w^+\,ct/w\,ct^+$. Among these 79 are red (bw^+/bw^+ : 2 bw^+/bw), 23 brown (bw/bw).

> All F_2 females have normal wings.

F_2 males: 86 white: $w\,ct^+/\!\leftarrow$

> 14 white, cut: $w\,ct/\!\leftarrow$

> 14 colored eyes ($w^+\,ct^+/\!\leftarrow$), 10 red and 4 brown

> 86 colored eyes, cut wing ($w^+\,ct/\!\leftarrow$), 65 red and 21 brown

The males display the recombination between white and cut, the first and last categories (86 each) being the parental, the remainder being recombinant. Only among the w^+ flies do we see, in each category, the segregation of 3 $bw^+\_\_$: 1 bw/bw.

2. The organism is haploid. The appearance of normal progeny of two slow-growing strains demonstrates that the two slow-growing strains do not carry allelic mutations. Therefore, one must symbolize the simplest interpretation of the cross as follows: $slo1^+\,slo2^- \times slo1^-\,slo2^+$, in which *slo1* and *slo2* are different genetic loci, each with wild-type and mutant alleles. Accordingly, the outcome of the cross gives us four genotypes, one of which almost certainly represents the 68 normal progeny:

> $slo1^+\,slo2^+$ (68)
>
> $slo1^-\,slo2^+$
>
> $slo1^+\,slo2^-$
>
> $slo1^-\,slo2^-$

Are the two genes independently assorting? If they were, we should see 1/4 normal and 3/4 slow-growing, assuming the double mutant is slow-growing. However, we see more like a 1/3 normal : 2/3 slow-growing. The "distortion" of the ratio cannot reflect linkage, because linkage would reduce the normal recombinants to *less* than 1/4. Therefore, we must suspect that some of the progeny are missing—they did not survive to become part of the progeny. Logically, suspicion falls on the double mutant class, which might not be able to grow at all. (Under this hypothesis, the genes appear to assort independently: there would be equal numbers of the surviving *slo* classes). That would leave a large majority or exclusively single mutants among the slow-growing progeny. To test this idea, one would mate a sample of the slow-growing progeny to each of the two parents. Single mutants would yield normal strains with one of the two parents, while double mutants will not yield normal progeny with either parent. In this way a prediction of the interpretation, the absence of double mutants, will be tested rigorously. One might also find that the germination of progeny spores is impaired, accounting for the missing class. (In Chapter 17, we will describe a second way of distinguishing different mutations with the same phenotype.)

3. This problem is somewhat different from the last. However, the haploid genotypes are the same as in the previous problem: $slo1^+ slo2^- \times slo1^- slo2^+$. Few as they are, the two normal progeny show that recombination between two slow-growing strains can reconstruct a normal strain, and the two mutations therefore cannot be strict alternatives, as alleles must be. The outcome of the cross gives us the same four genotypes, but here there are only two $slo1^+ slo2^+$ genotypes.

$$slo1^+ \ slo2^+ \ (2)$$
$$slo1^- \ slo2^+$$
$$slo1^+ \ slo2^-$$
$$slo1^- \ slo2^-$$

The single- and double-mutant progeny account for the rest. Here, we have no reason to assume that the double mutant is a lethal gene combination; there is no evidence for this in the ratio of progeny. (They would account for only about 2 of the 443 slow-growing progeny or 0.5%.) The initial observations are a powerful indication of linkage, and to estimate the map distance, one must recognize that recombinants in this progeny of 445 are not only the 2 that are visible, but probably about 4, to account for the double mutants. The map distance would be, therefore, about 0.9 map unit. The most rigorous follow-up test, assuming one has other genetic markers available, is to find linkage of one of the mutants (e.g., $slo1^-$) to an unrelated gene and then see whether $slo2^-$ is also linked to that gene at a distance compatible with the 0.9 map units between *slo-1* and *slo-2*.

4. This clearly throws suspicion on the Y chromosome as the bearer of the mutation in question. The uniformity of phenotype among males, with transmission from father to sons, cannot be accounted for by autosomal or X-linked inheritance. This trait, determined by a presumed Y-linked gene, is in fact known.

5. The organism is haploid. Strains #1 and #2 clearly differ: the first having a single allelic difference with the His⁻ strain (his^+ vs. his^-); and the second having at least two, one of which is the difference at the known *his* gene. The simplest interpretation of the cross with Strain #2 is that another gene can "suppress" the expression of the his^- allele, either by providing a duplicate function or by bypassing or mitigating the effect of the his^- mutation in some other way. If we call this gene *hsu* ("histidine mutant suppressor), with the active allele of the gene (hsu^+) in Strain #2, and the inactive allele (hsu^-) in the His⁻ strain, the second cross would be symbolized:

$$his^+ \ hsu^+ \times his^- \ hsu^-,$$

yielding the following progeny:

$his^+\ hsu^+$ (His$^+$)

$his^-\ hsu^-$ (His$^-$)

$his^-\ hsu^+$ (His$^+$)

$his^+\ hsu^-$ (His$^+$)

If the genes assorted independently, the His$^+$: His$^-$ ratio would be 3:1. The hypothesis can be tested by mating each of the His$^+$ progeny to strain #1. One-third of them should yield 1/4 His$^-$ progeny. Of the His$^+$ progeny that do not do so, they should be equally divided between those resembling Strain #1 and those resembling Strain #2 when mated again with a His$^-$ strain.

You might be tempted to interpret the results of the original cross of the His$^-$ strain and Strain #2 in another way: that Strain #2 has a mutation which when combined with the his^- mutation, renders the double mutant lethal. However, this would yield a ratio of 2 His$^+$: 1 His$^-$, the maximum ratio permitted by independent assortment. In contrast, the actual outcome is a 3:1 ratio.

6. *Drosophila* is diploid, of course, but the bald trait is peculiar. It appears only among males of a true-breeding strain; at the outset it appears to be a trait that is strictly sex-limited in its expression. The first cross indicates that it is not only sex-limited in expression, but sex-linked as well, the F$_1$ and F$_2$ progeny satisfying one criterion of a single, sex-linked gene. The cross should be diagrammed with good notation beginning at this point, using the facts of the first cross. This will allow you to expand upon it when you interpret the second cross.

Cross 1:

Parents bd^+/bd^+ females $\times$ $bd^-/\leftarrow$ (bald) males

F$_1$: bd^+/bd^- females and $bd^+/\leftarrow$ males, both normal

F$_2$: bd^+/bd^+ and bd^+/bd^- females, both normal

$bd^+/\leftarrow$ (normal) and $bd^-/\leftarrow$ (bald) males in equal numbers

Cross 2 shows exactly the same results, with one exception: the appearance of 1/8 bald females. The first step in resolving this paradox is to diagram the reciprocal of the Cross 1 because Cross 2 is exactly that. Accordingly:

Cross 2 (preliminary):

Parents bd^-/bd^- females (**why not bald?**) $\times$ $bd^+/\leftarrow$ males

F$_1$: bd^+/bd^- females (normal) and $bd^-/\leftarrow$ (bald) males

F$_2$: bd^+/bd^- and bd^-/bd^- females (**why only 1/8?**)

$bd^+/\leftarrow$ (normal) and $bd^-/\leftarrow$ (bald) males in equal numbers

In this cross, the questions in bold letters signify the discrepancy between the prediction and the actual results observed. If we look at this closely, we see that the parental females do not express the bald phenotype (nor do they ever do so in the true-breeding line), but some contribution of the other strain allows some F$_2$ females to do so. Therefore, we need to look for a genetic factor that explains this and to make genotypes consistent with all other data. If baldness were expressed in homozygous females, it would appear in parental females and in 1/2 of the F_2 females. However, only 1/4 of the expected value is obtained. This is the criterion ratio of a single, autosomal recessive in an F_2, in which original, true-breeding parents differed.

The two stocks appear to differ in an autosomal recessive gene that, when homozygous, allows expression of the bd^- mutation in females. This factor is not required for expression of bd^- in males. The recessive gene (which we will call a) is found in the stock having no baldness; clearly it cannot impart baldness by itself. Thus the original stocks were: Stock #1: $a/a\ bd^+/bd^+$ females and $a/a\ bd^+/\leftarrow$ males; Stock #2: $A/A\ bd^-/bd^-$ females and $A/A\ bd^-/\leftarrow$ (bald) males.

CHAPTER 12
ANSWERS TO PROBLEMS

1. The frequency of a trait in a population has no necessary relationship to its dominance or recessiveness. A common trait may be recessive.

2. In this population, gametes are produced according to the frequencies of alleles in the population: $f(A) = 0.5$ and $f(a) = 0.5$. Using the multiplication law, $f(AA) = 0.5 \times 0.5 = 0.25$; $f(Aa) = 0.50$ (arising from two gamete combinations, A sperm $+ a$ eggs, and a sperm $+ A$ eggs, each with a frequency of 0.25); and $f(aa) = 0.5 \times 0.5 = 0.25$. This is equivalent to the binomial expansion, or $(p + q)^2$, the algebraic formulation of the Hardy-Weinberg equilibrium. After one generation of random mating, $f(A) = f(AA) + 1/2$ $f(Aa)$, or 0.5, and $f(a) = f(aa) + 1/2f(Aa)$, or 0.5. The allele frequencies are unchanged, as are the genotype frequencies in the second generation of random mating. An equilibrium has therefore been achieved in the first generation.

3. The key to solving this problem is to first determine the frequency of the recessive category. It is $1 - 0.75 = 0.25$. At Hardy-Weinberg equilibrium, $q^2 = f(aa) = 0.25$, and therefore $q = f(a) = 0.5$. By subtraction, $p = f(A) = 0.5$. Only because you know that the population is in Hardy-Weinberg equilibrium can you use the frequency of aa homozygotes to represent the term q^2. Organisms having the dominant phenotype will include both AA and Aa genotypes, in initially unknown proportions. Therefore, $q\ [= f(a)]$ cannot be known from the frequency of aa homozygotes without further information.

4. There is sufficient information to calculate allele frequencies: $f(A) = 0.05 + 1/2(0.20) = 0.15$ and $f(a) = 1/2\ (0.20) + 0.75 = 0.85$. Allele frequencies are substituted into the Hardy-Weinberg formula, and expected genotypic values are calculated: $f(AA) = 0.0225$, $f(Aa) = 0.255$, $f(aa) = 0.7225$. These values are close to, though not identical with, the observed genotypic proportions. Because we are dealing with proportions, rather than actual numbers, we cannot perform a χ^2 test. Note that it would be incorrect to equate the genotypic proportions with p^2, $2pq$, and q^2, because there is no certainty that this population is in Hardy-Weinberg equilibrium.

5. In this population, $f(N) = 0.304$ and $f(R) = 0.696$. Substituting these numbers into the Hardy-Weinberg formula gives $(0.304)^2 = 0.0924$ (the frequency of NN) + $2(0.304 \times 0.696) = 0.423$ (the frequency of NR) + $(0.696)^2 = 0.484$ (the frequency of RR). To calculate the expected number of cattle in each category at Hardy-Weinberg equilibrium, multiply the frequencies by the total number of cattle in the population (8705). The expected numbers are 804.48 NN, 3683.68 NR, and 4216.84 RR.

6. The observed number of cattle in each category was 756 NN, 3780 NR, and 4169 RR. A chi-squared analysis (*see* Chapter 5) is used to determine whether the observed and expected numbers are significantly different from one another. Chi-squared is calculated as the sum of the squared deviations (observed minus expected), each divided by the

expected values. For the cattle example, we substitute our observed and expected values: $\chi^2 = (756 - 804.48)^2/804.48 + (3780 - 3683.68)^2/3683.68 + (4169 - 4216.84)^2/4216.84 = 5.983$. For this problem, we have one degree of freedom. We begin with three genotypic classes but subtract one degree of freedom because establishing any two genotypic values automatically fixes the third. We must subtract an additional degree of freedom because the data are reduced to the p and q values, within which there is only one degree of freedom. Our χ^2 value of 5.983, with one degree of freedom, yields a P value between 0.05 and 0.01). We conclude that the observed values deviate significantly from the expected values; the population is not in Hardy-Weinberg equilibrium.

7. The population might be in Hardy-Weinberg equilibrium, but selection could also explain the observed genotypic frequencies. This would require looking at successive generations. To illustrate this, consider a population having the following frequencies: $f(BB) = 9/25$, $f(Bb) = 12/25$, $f(bb) = 4/25$. These numbers represent the expansion of the binomial $(3/5 + 2/5)^2$. Assume that at the time of zygote formation, half of the heterozygotes and three quarters of the bb homozygotes are lost to selection. This leaves 9/25 BB homozygotes, 6/25 Bb heterozygotes, and 1/25 bb homozygotes, for a total of 16/25 (9/25 of the population being lost to selection). To normalize the frequencies to 1.0, we divide genotypic frequencies by the new total, 16/25, and obtain the following adult relative frequencies: $f(BB) = 9/16$, $f(Bb) = 6/16$, and $f(bb) = 1/16$. These values sum to 1.0, and conform to an expansion of the binomial $(3/4 + 1/4)^2$. Thus even though the initial population conformed to the Hardy-Weinberg equilibrium, gene frequencies were altered by selection in the next generation. Technically, information on genotype frequencies at several points in the life cycle, or over several generations would be necessary to establish that a true equilibrium existed.

8. (i) The allele frequency of a recessive allele is equal to the square root of the frequency of homozygotes in the population: the square root of 0.0016 is 0.04. This approach assumes Hardy-Weinberg equilibrium, which is appropriate for severely deleterious alleles when they are rare. (ii) $1 - 0.04 = 0.96$. (iii) $2 \times 0.96 \times 0.04 = 0.0768$.

9. (i) Genotype frequencies: $FF = 30/50 = 0.6$, $FS = 6/50 = 0.12$, $SS = 14/50 = 0.28$. (ii) Allele frequencies: There are 100 alleles in the population. The FF individuals have 60 F alleles, and FS individuals have 6 F, for a total of 66 F alleles, or 0.66 of the total. Similarly, there are a total of 34 S alleles (6 + 28), or 0.34 of the total. (iii) If $f(F) = p$ and $f(S) = q$, the Hardy-Weinberg law yields the expected genotype frequencies: $f(FF) = (0.66)^2 = 0.4356$, $f(FS) = 2 \times (0.66)(.34) = 0.4488$, and $f(SS) = (.34)^2 = 0.1156$. (iv) The inbreeding coefficient, F, is a measure of the deviation of H_1, the observed frequency of heterozygotes from that expected in a population in Hardy-Weinberg equilibrium: $F = (2pq - H_1)/2pq$. In this case, $F = (0.4488 - 0.12)/0.4488 = 0.733$.

10. (i) The frequency of an X-linked recessive allele (cb) is equal to the frequency of males displaying the phenotype; in this case, $f(cb) = 0.012$. The frequency of cb^+ alleles is $1 - 0.012 = 0.988$. (ii) The frequency of heterozygous carriers (among females) is $2(0.012)(0.988) = 0.024$ (2.4%). (iii) The expected frequency of color-blind females is $(0.012)^2 = 0.000144$ (0.0144%).

11. Genetic drift is an important evolutionary force that operates in small populations. Drift leads to increases in homozygosity in smaller populations and may fix particular alleles in some of them. The probability of mating among close relatives rises in isolated populations, increasing the probability of homozygosity for deleterious, recessive alleles.

12. (i) $(0.01)^2 = 0.0001$; in a population of 1,000,000, 100 will lack chlorophyll. (ii) The lethal recessive allele will persist for many generations at low frequencies, because most are in heterozygotes, sheltered by the dominant allele.

13. (i) This is an example of heterozygote superiority. (ii) Allele frequencies remain at equilibrium. (iii) Because selection eliminates particular genotypes at some stage in the life cycle, the populations will not be at Hardy-Weinberg equilibrium.

14. (i) The chasmogamous flowers provide opportunities for outcrossing, which helps maintain genetic diversity in a population. The cleistogamous flowers generally guarantee pollination success, and the inbreeding associated with self-pollination may help a population adapt to specific conditions. (ii) A population with pollination success only in cleistogamous flowers would show an excess of homozygotes relative to expectations based on the Hardy-Weinberg principle. A population with pollination success only in chasmogamous flowers may mate randomly and should have more heterozygotes than the other population.

15. (i) Population I: 3.2-kb allele: 0.7; 4.0-kb allele: 0.3. Population II: 3.2-kb allele: 0.2; 3.8-kb = 0.8. (ii) Both populations appear to be in Hardy-Weinberg equilibrium. (iii) Migration would introduce the 3.8-kb allele into population I and would change the relative frequencies of all alleles.

16. The number of different types of gametes in the male $= 2^4 = 16$. Intrachromosomal recombination in females gives rise to thousands of additional gametic types; the number depends upon the frequency of crossing-over.

17. Although most mutations are deleterious, some are neutral with respect to natural selection, and a few prove to be advantageous in specific environments. Without new mutations, there is nothing for any of the known evolutionary forces (natural selection, genetic drift, etc.) to act upon to lead to adaptation of natural populations to their environment.

18. One experiment to determine the relative fitnesses of the different alleles would be to release the moths into environments with different substrates. The release of known numbers of artificially-marked light and dark moths into environments where the substrates are either light or dark, offering camouflage to one or the other phenotype, allows measurement of relative mortality in a recapture experiment. In all likelihood, the selection coefficients will differ in the different environments, with dark moths having a higher selection coefficient (lower fitness) in the light area, and the light allele having a higher selection coefficient in the dark areas. The dark allele should be eliminated more rapidly in light environments because it is dominant.

19. The expected genotype frequencies follow the Hardy-Weinberg principle expanded for three alleles: $p^2 + 2pq + 2pr + q^2 + 2qr + r^2 = 1$. If $f(I^A) = p$, $f(I^B) = q$, and $f(I^O) = r$, then the expected frequency of each phenotypic class is as follows: type A $= f(I^A I^A) + f(I^A I^O) = 0.242$; type B $= f(I^B I^B) + f(I^B I^O) = 0.203$; type O $= f(I^O I^O) = 0.52$, and type AB $= f(I^A I^B) = 0.039$. Multiplying each of these by 100 and rounding to the nearest whole number yields: 24 type A, 20 type B, 52 type O, and 4 type AB.

20. (i) Locus I: f(3.6 kb) = 0.4, f(2.8 kb) = 0.6. Locus II: f(4.0 kb) = 0.44, f (3.3 kb) = 0.56. (ii) Locus I is probably at Hardy-Weinberg equilibrium; for locus II, the expected frequencies based on the Hardy-Weinberg law are: 4.0-kb homozygotes = 0.1936; 3.2-kb homozygotes = 0.3136; heterozygotes = 0.4928. (iii) In this example, locus II shows a deficiency of heterozygotes, possibly as a result of selection against heterozygotes or non-random mating (matings among organisms having the same phenotype and presumably the same genotype for the locus in question). Another possibility is that there is migration into the population from another population where the allele frequencies are the same at locus I but different at locus II.

CHAPTER 13
ANSWERS TO PROBLEMS

1. Mean = (175 + 202 + 168 + 154 + . . .)/10 = 185.1 grams. Variance = $\Sigma(x - \bar{x})^2/(n - 1)$ = $(175 - 185.1)^2 + (202 - 185.1)^2 + (168 - 185.1)^2 + (154 - 185.1)^2 + (188 -$

$185.1)^2 + \ldots /(10 - 1) = 274.76$. Standard deviation is the square root of the variance or 16.58 grams.

2. The traits with the higher heritabilities are more likely to show resemblance among family members. Therefore height and weight are more likely to "run in families" than is overall fertility. The blood-pressure heritabilities suggest that there is a strong genetic component to this trait, and you should be concerned about blood pressure if your family has a history of high blood pressure.

3. The formula that is needed to solve this problem is: $M' = M + h^2(M^* - M)$, in which M is the mean percent milk protein in the herd, M^* is the percent milk protein in the cows selected for breeding, and M' is the percent milk protein in the offspring. In this case, $M' = 0.12 + 0.55 (0.18 - 0.12) = 0.153$. The offspring from the first set of crosses have 15.3% protein in their milk. To determine how many generations it will take to obtain milk with more than 17.5% protein, it is easiest simply to repeat the foregoing calculation, remembering to substitute the new mean percent protein at each generation. For the second generation, $M' = 0.153 + 0.55 (0.18 - 0.153) = 0.168$. The second generation of selection yields cows with 16.8% protein. Two more generations of selection after that yield herds with 17.5% and 17.8% protein, respectively. Note that the magnitude of the response to selection decreases with each generation.

4. The environmental variance for oil content is the variance of the F_1 generation. The variance of the F_2 generation is the total variance for the trait in the population. The genotypic variance is the total variance minus the phenotypic variance. The broad sense heritability is the genotypic variance divided by the total variance.

   F_1: Mean $= 8.2$; variance $= [(2.5 - 8.2)^2 (0) + (3.5 - 8.2)^2 (20) + (4.5 - 8.2)^2 (40) + (5.5 - 8.2)^2 (90) + \ldots]/(999) = 3.85$. F_2: Mean $= 7.49$; variance $= 7.55$.

   Environmental variance $= 3.85$; genotypic variance $= 7.55 - 3.85 = 3.7$; broad-sense heritability $= 0.49$.

5. The fact that selection for the number of eggs per hen is difficult suggests that there is little variation for this trait among chickens or the fowl from which they were domesticated. The heritability for this trait must be relatively low.

6. Environmental variance $= 12.6$; total variance $= 20.3$; genotypic variance $=$ total variance $-$ environmental variance $= 7.7$. Broad-sense heritability $=$ genotypic variance/total variance $= 0.38$.

7. The relevant formula for this problem is $M' = M + h^2(M^* - M)$. Solving for $h^2 = (M' - M)/(M^* - M)$ where M' is the mean height of the progeny from the cross, M^* is the height of the individuals selected for breeding, and M is the mean height of the initial population. For the mowed field: $h^2 = (6.5 - 6.2)/(8.0 - 6.2) = 0.167$; for the unmowed field, $h^2 = 0.32$.

   The heritability, broad-sense or narrow-sense, is a function of the population or populations under consideration. It should not be a surprise that the heritability of this trait is different in the two populations. One reason for the difference could be that the population in the mowed field was started by a few individuals with initially low genetic variability. Another possibility is that repeated mowing selects for individuals that flower rapidly between mowing, which may have reduced the genetic variability and hence the heritability of this trait in the plants from the mowed field. It is not uncommon for traits endowing overall fitness so directly to have low narrow-sense heritabilities.

8. Development of a new breed of any plant or animal frequently includes inbreeding, and this may expose rare recessive alleles in the gene pool. This is a special problem when breeders start with only a few animals and create entire stocks from their descendants. Correlated responses to selection also contribute to the development of undesirable char-

acteristics when selecting for desirable characteristics. Selection for a nicely shaped head or a particular leg length in a dog may have undesirable effects on other skeletal features. Genetic linkage of desirable alleles at some loci and undesirable alleles at other loci contribute greatly to the correlated selection of two or more different traits. There are usually limits to selection, so twelve-foot dachshunds will probably never be seen.

9. The traits with the higher heritabilities, such as percent protein in milk, would be most responsive to selection. The calving interval would be least responsive to selection.

10. The levels of heterozygosity in the first two strains of corn would decrease in the course of the inbreeding. The inbred strains would probably yield fewer bushels of grain per acre than the strains from which they were derived. This phenomenon is referred to as inbreeding depression. The third stain of corn is called a hybrid strain. This strain would probably yield more grain per acre than the two inbred strains. Hybrid vigor, or heterosis, is the term used to describe the enhanced production of the hybrid corn.

11. Restriction fragment length polymorphisms (RFLP's) can be used to help understand the genetic basis of quantitative traits. RFLP's represent differences in the DNA at a specific genetic locus. The differences, or polymorphisms, are due either to mutations in the DNA sequence that the restriction enzyme recognizes or to insertions or deletions of DNA at or near the locus of interest. It is possible to test for associations between different RFLP alleles at several loci and phenotypic measurements such as blood pressure. To do this, it is necessary to have cloned DNA segments representing the RFLP loci of interest, and DNA samples from many individuals. Associations between a specific RFLP and a specific phenotype suggest that a genetic locus that may influence the phenotype is physically linked to the RFLP locus.

12. Phenotype is the result of interaction between gene and environment. The genetically identical fields of wheat differ phenotypically because the environments in which they were grown are different. The specific crops that were grown in the preceding years must have affected the soil. In this case, it is likely that the nitrogen added to the soil by the soybeans increased the growth rate and yield of the wheat.

CHAPTER 14
ANSWERS TO REVIEW QUESTIONS

1. (i) Maternal effect: the maternal genotype *may* override the genotype of the offspring, but the phenotype does not persist without the appropriate nuclear gene in succeeding generations. Where reciprocal crosses yield *uniformly* different phenotypes (in both sexes), one must pursue the observation by further testcrosses to establish that the anomaly is determined by the maternal *genotype*. (ii) Cytoplasmic inheritance: in reciprocal crosses, the appearance of the (usually) maternal phenotype in all offspring, and the lack of segregation of differences of the trait in testcross progeny or F_2 progeny of such offspring. The trait should persist in later generations. (iii) Sex linkage: in one of the reciprocal crosses of homozygous stocks, a difference between the sexes in the distribution of a trait, and a departure, in both sexes of the F_2 from a 3:1 or 1:2:1 ratio, which are characteristic of autosomal inheritance. Segregation, however, is observed.

2. All F_1s (*Aa*) would be red whether the mother was *AA* or *aa;* all the F_2s (*AA, Aa,* and *aa*) would be red, owing to the genotype of the F_1 mothers. Testcrosses of the F_1 males with an *aa* female would yield a ratio of 1 red : 1 white; testcrosses of the F_1s with an *aa* male would yield all red larvae. Among the F_2s, testcrossing males with *aa* females would give a ratio of 1/4 that yielded only red, 1/2 that yielded 1 red : 1 white, and 1/4 that yielded only white progeny. Testcrossing F_2 females with an *aa* male would give 3/4 (the *A_* class) that yielded only red larvae, and 1/4 (the *aa* class) that yielded only white.

3. A cross of the two petites would yield a normal diploid. Its haploid meiotic products would segregate 1 normal (pet^+) : 1 petite (pet^-). The cytoplasmic mutant character would have disappeared from the lineage in the formation of the diploid.

4. Testcrosses of the F_1, or intercrosses of the F_1 flies to form an F_2 is necessary to have flies in which segregation of genotypes might occur. This segregation, however, might not be visible phenotypically, given the results of the reciprocal crosses. Because the parents are true-breeding, it would be possible to perform reciprocal crosses of the F_1 of *both* crosses with *both* parental types. In doing this, we free ourselves of any presupposition about a nuclear or a cytoplasmic origin of the small phenotype. If Mendelian segregation failed to occur in any "test" cross, and if the phenotype of the offspring always resembled the maternal parent, a good case for cytoplasmic inheritance is established.

5. Organelles are indispensable entities in cells and have no independent, free-living state. Few of their proteins are encoded in their DNA, which is relatively small. Symbionts, though they may not be free-living, have all or most of the genes needed for macromolecular synthesis and can duplicate with many fewer contributions from the nucleus of the host. In general, symbionts are not essential to the host. (Mitochondria and chloroplasts have recently been called symbionts, but this is simply to recognize their origin, not their present relationship with the host.)

6. The slow-growing character is probably a cytoplasmic determinant that asserts itself in the diploid and affects all meiotic products (it does not segregate). Only among these slow-growing progeny do we see segregation for a uracil requirement. Its failure to appear in the slow-growing diploid, and its appearance in the slow-growing haploids, suggests that it is recessive, and the segregation tells us that this requirement is probably determined by a nuclear gene. The normal phenotype of the normal parental strain (haploid) tells us that the uracil requirement cannot be displayed unless the slow-growing determinant is also present.

CHAPTER 15
ANSWERS TO REVIEW QUESTIONS

1. Transcription: the template is DNA; it is read by RNA polymerase; and the product is RNA. Translation: the template is mRNA; it is read by ribosomes and amino-acyl-tRNA; and the product is a polypeptide chain. The processes differ in that transcription reads DNA sequences of dA, dG, dC, and dT into equivalent RNA sequences of rA, rC, rG, and rU on a nucleotide-for-nucleotide basis. Translation decodes the 64 RNA nucleotide triplets into twenty protein amino acids. Because three mRNA nucleotides in sequence specify one amino acid, the coding ratio is 3. The overall lengths of mRNA sequences and their corresponding polypeptides therefore differ.

2. Transcription requires (i) DNA, the source of information; (ii) ribonucleotide triphosphates, the raw materials; and (iii) RNA polymerase, the enzyme that catalyzes the linking of ribonucleotides into RNA chains. RNA polymerase binds to a specific sequence on DNA near the point at which transcription must begin; the DNA chains separate; and RNA polymerase copies one of the DNA chains into a complementary RNA chain. This process uses the base-pairing rules of DNA replication, except that in the RNA chain the complement of adenine in the template DNA is uracil, rather than thymine. Transcription of the RNA chain proceeds from its 5′ to its 3′ end, using the DNA chain of opposite polarity.

3. Amino acyl-tRNA synthetases are enzymes that simultaneously recognize individual tRNAs and the cognate amino acids and link them together. This is the point at which the chemically dissimilar building blocks of protein (amino acids) and nucleic acids (the nucleotides of the anticodon of the tRNA) are joined together. In formal terms, translation is

being done here, with the tRNA synthetases and their products (amino acyl-rRNAs) acting as a dictionary. The later codon-anticodon recognition on the ribosome may be
thought of as an extension of the translation process that began with the amino acyl-
tRNA synthetase reactions.

4. Translation requires (i) mRNA, which bears genetic information from DNA; (ii) aminoacyl-
    tRNAs, a collection of small tRNA molecules to which amino acids have been attached at
    one end by aminoacyl-tRNA synthetases (each amino acid corresponding to the anticodon
    of the particular tRNA); (iii) ribosomes, aggregates of ribosomal proteins and rRNAs, on
    the surface of which the mRNA and aminoacyl-tRNA's engage in codon-anticodon pairing.
    The ribosomes, together with additional protein factors, are responsible for stabilizing the
    interactions, relational movement between mRNA and the ribosome, and certain chemical
    reactions in the formation of peptide bonds.

5. (iii): There are three bases in sequence for each amino acid.

6. (i) Messenger RNA (mRNA), the carrier of information; (ii) transfer RNA (tRNA), the
    adaptor for amino acids during the reading of mRNA; and (iii) ribosomal RNA (rRNA), a
    structural component of ribosomes and catalytic participant in some reactions of transla-
    tion. All types of RNA are encoded in corresponding genes. Most genes in the nucleus
    specify mRNA. There are about 40 genes, some of them in multiple copies, for tRNAs.
    Only four types of genes (in multiple copies in eukaryotes) encode the four rRNAs.

7. (i) 5'–3'; (ii) 5'–3'; (iii) N-terminal to C-terminal; (iv) 3'–5'.

8. mRNAs differ from the corresponding primary transcript by having a 5' cap (7-methyl
    guanidine), a 3' poly(A) tail, and no introns.

CHAPTER 16
ANSWERS TO REVIEW QUESTIONS

1. (a) Base substitution (transitions and transversions): replacement of one base pair by an-
    other (e.g., AT by GC). This may change the meaning of the code at that position, leading to
    amino acid substitution or polypeptide chain termination in translation of the corresponding
    mRNA. (b) Frameshift: addition or deletion of 1–4 base-pairs. Except for changes of three
    nucleotides, this will lead to resetting the reading frame of a gene, which wholly disrupts
    the translation of the mRNA coding region. (Additions or deletions of three contiguous nu-
    cleotides are not really frameshifts. However, because they are rarely detected, they remain
    included in this category of mutation by virtue of the way they arise.) (c) Macrolesions:
    larger-scale alterations of DNA such as large deletions, insertions, translocations, inversions
    and changes in ploidy (euploid or aneuploid).

2. Mutation is a change in the information in DNA which is nevertheless chemically normal.
    DNA damage is chemically abnormal DNA, which in many cases makes it recognizable
    by repair enzymes and susceptible to repair.

3. UV causes mutation by inducing pyrimidine dimers in one of the two nucleotide chains
    of DNA. These cause mutations if they are repaired or replicated by the error-prone SOS
    repair system, in which replication inserts erroneous bases opposite the dimer. UV can
    also indirectly cause chromosome breaks, which lead to macrolesions. Pyrimidine dimers
    can be repaired by the photoreactivating enzyme (which does not involve excision), and
    by excision repair, in which the dimers and nearby bases are removed and replaced by an
    accurate repair polymerase. Both processes are virtually error-free. Recombination repair
    defers the serious effect of pyrimidine dimers on DNA replication, allowing the other re-
    pair mechanisms time to act upon the dimers.

4. Damage (broadly defined) that does not involve the occurrence of abnormal nucleotides includes frameshifts and mismatched bases. In both cases, these can be repaired at the heteroduplex stage by enzymes that recognize and correct mismatches. The accuracy of repair will not be the same as in the case of damaged bases, because the original and the damaged DNA chain are not usually distinguished by repair systems. Among the types of larger scale damage (macrolesions), few can be repaired easily, although the tendency of broken chromosome ends to join together greatly minimizes the potential damage and lethality.

5. "Mutator genes" have been found in most genetically well-studied organisms. Among the functions affected are the enzymes of DNA synthesis, which might have impaired fidelity of replication; the enzymes of the repair systems, as in *xeroderma pigmentosum;* enzymes that make, or fail to neutralize, mutagenic compounds such as base analogs or free-radicals.

6. Two major processes, not one, act in the evolution and adaptation of organisms to their environment. The first is a chance process: mutation, occurring randomly in populations. The second is also a "blind," but decidedly non-random process, the differential effect of a particular environment upon reproductive rates of organisms carrying different alleles of a relevant gene. Mutational variants that confer a better chance of reproduction (or even survival) in that environment will, over several or many generations, become more prevalent, often to the point that the original allele of the gene disappears entirely.

   Your friend—or even you—may not be wholly convinced by this explanation, but at least you can point to a non-random process (natural selection) that automatically adapts a population of organisms to an environment if suitable mutational variability exists. Chance mutation is only a component of the process, but not sufficient for adaptation. It should impress us, as Jaques Monod has said, that these blind processes, mutation and natural selection, could give us vision itself.

CHAPTER 17
ANSWERS TO QUESTIONS AND PROBLEMS

1. In *Neurospora,* recombination is impossible because the genes involved are in different nuclei of vegetative cells. In yeast and *Drosophila,* the genes are in diploid nuclei, but the cells (or organisms) that are tested phenotypically are vegetative entities that are not undergoing chromosomal recombination.

2. If strains X and Y carry allelic mutations that display intragenic complementation, then Z is a one-gene mutation that does not complement with either of them. A test of this hypothesis would be a cross of strain Z to the wild-type strain. The cross should yield a 1:1 segregation of His$^+$ and His$^-$, and a cross of strains X and Y should yield few or no wild-type recombinants.

   If, on the other hand, strains X and Y carry non-allelic mutations, then strain Z is a double mutant carrying mutations in both genes. Thus the cross of strains X and Y should with high probability yield wild-type progeny at frequencies sufficient ($> 0.5\%$) to indicate recombination of non-allelic genes. A cross of strain Z and wild type should yield 25% wild-type progeny if the component mutations are unlinked, and more if they are linked. The mutant progeny of this cross should be classifiable in complementation tests as one or the other single mutant, or the double mutant, using strains X and Y as testers. If two single mutants were to be found in the progeny of this cross, it would demonstrate the double-mutant character of the Z parent.

3. Complementation tests are done by introducing two independently isolated mutations into vegetative diploid or heterokaryotic cells in the *trans* arrangement. If this double het-

erozygote or heterokaryon has a normal phenotype, it reveals that the mutations are different from one another as well as recessive to their respective wild-type alleles. Each haploid contribution to the diploid or heterokaryon clearly has an intact, functional wild-type allele of the mutation carried by the other haploid genome. The intact wild-type allele should be thought of as the whole segment of DNA that encodes a polypeptide, and therefore has a long sequence of mutable sites. No recombination takes place, or is allowed, in this test, just a test of independent function of the wild-type alleles.

Recombination tests bring two mutant, haploid genomes together in a diploid cell that will undergo meiosis, in which recombination of genes and chromosomes is expected. The meiotic products of the mating are haploid, and the test asks whether recombination can reconstruct a normal *haploid genotype* from the two mutant parental nuclei. Genes that are non-allelic will be able to recombine, although it may occasionally be at a low frequency because of tight linkage of the mutations. However, allelic (non-complementing) mutations will also be able, theoretically, to recombine at a very low frequency if the sites of mutation in the gene in question are different. Therefore, recombination tests resolve mutational sites in the same or different genes; they do not rigorously test allelism.

4.   The auxotroph is a double mutant, constituted of two independently assorting histidine mutations. A mating with wild type yields only 25% of the progeny carrying neither mutation.

5.   The mutations a and e represent different genes; a third gene is represented by the other mutations. A complex pattern of intragenic complementation is seen, but mutation b is a non-complementer embracing all of its alleles (c, d, f, g).

6.   a. Four genes are involved, having the alleles as follows: (1) a, c, f, g, k, l; (2) b; (3) d, h; (4) $e, i, j. i$ and j show intragenic complementation.

     b.

$$\underset{X}{} \xrightarrow{f} B \xrightarrow{h} C \xrightarrow{e} A \xrightarrow{b} \text{zotobiose}$$

     c.

$$f \text{——} 6 \text{——} (i,j) \text{——} 8 \text{——} b \quad (h \text{ assorts independently})$$

     d. The 3% prototrophs imply an equal frequency of double mutants, yielding 6% recombinant meiotic products. These require that 12% of the tetrads must be tetratypes (in which only half the meiotic products are recombinant). No NPDs form if there are no double crossovers.

     e. The missing classes are the doubles (HRt and hrT), which can be used to infer gene order. The map is R—3—H—2—T.

7.   At the outset, you will need to predict the phenotype of a mutant unable to make a normal cell wall. If the cell wall contains the pressure (turgor) of the cell, we would predict that cells would be sensitive to low osmotic pressure in the medium. Therefore, we would look for mutations that grew well in the presence of high concentrations of, say NaCl or sucrose, but died or grew poorly when put in normal or low-osmotic media. The next step is to devise a method for obtaining mutants with this phenotype. After irradiation of a population of wildtype yeast, you might grow the survivors by plating them on a medium with high osmolyte concentrations, then replica-plate them to normal media. Those that failed to grow on the latter media would be revealed, and isolated (always to a medium

permissive of growth). Genetic analysis would follow according to the methods in this chapter. The various mutant classes, identified by complementation tests, should then be inspected closely (by chemical and cytological means) for any phenotypic distinctions, such as what specific components of the cell wall were altered or missing. If certain genes eliminated all cell-wall polymers, for example, while others eliminated only individual ones, an order or dependence of function among genes could be inferred, and tested in double mutants.

CHAPTER 18
ANSWERS TO QUESTIONS

1. Conjugation starts with the formation of a conjugation bridge and the entry of part of the F DNA, followed by variable amounts of the bacterial chromosome attached to it during rolling-circle replication. The conjugation bridge breaks spontaneously at different times for different cell pairs, and the transferred DNA is then integrated by recombining with the recipient chromosome. The limiting factor in acquisition of a gene is its position relative to the origin of transfer (that is, to the position of the F factor in the Hfr). Both the initiation of the conjugation process and the chromosomal integration of an Hfr gene, once inside the F^- cell, are quite efficient.

2. Selective markers are used to prevent growth of the many parental cells in a conjugation mixture. Recombinants, which may be quite rare for markers transferred late in conjugation, or those arising in transduction, would be swamped by the many parental cells.

3. Late genes are transferred with low efficiency because few conjugating pairs remain joined by a bridge by the time a late gene might enter F^- recipient cells. The exconjugant F^- cells remain F^- because only a portion of the F factor enters at first. In the rolling-circle replication that accompanies chromosome transfer, however, a copy of the entire F factor enters the F^- cell after 100 minutes, and then rare Hfr exconjugants can be found.

4. Rare transfer of chromosomal genes between F^+ and F^- cells takes place in two steps: (i) the very rare formation of an Hfr from an F^+ cell by integration of the F factor, and (ii) conjugation and chromosomal transfer between this Hfr cell and an F^- cell.

5. By integration of the circular F plasmid with a single recombination event, into the circular bacterial chromosome.

6. Injection of DNA by the phage coat into the bacterial cell is followed by (i) cessation of bacterial macromolecular synthesis, often accompanied by breakdown of the bacterial chromosome; (ii) the appearance of metabolic activities required in phage replication, catalyzed by enzymes encoded in the genes of the phage (iii) replication of phage DNA; (iv) synthesis of phage coat proteins and enzymes that weaken the bacterial cell membranes; (v) encapsidation of the phage DNA by phage coats; and (vi) lysis of the host, liberating the phage.

7. Plasmid: a circular DNA molecule in a cell that is in general dispensable to the cell. These DNA's in some cases have other forms, as in the DNA of P1 phage particles, or the integrated form of the F factor. Temperate bacteriophage: a bacteriophage having two possible fates after injection into a host: it can lysogenize the host by entering the prophage state, or it can undergo the lytic cycle, leading to the formation of many more phage particles. Virulent bacteriophage: a phage that always undergoes the lytic cycle upon infecting bacterial cells.

8. They must be close to one another on the bacterial chromosome. In the case of the common transducing phage P1 of *E. coli*, the markers must be within 2% of the length of the bacterial chromosome (two minutes, in terms of the interrupted-mating map).

9. Because it is unlikely that two genes drawn at random would be so closely linked that they would be cotransduced. Cotransduction is a minimal requirement of mapping genes by this method. If they are too far apart, the linkage in the chromosome is automatically broken, and therefore unmeasurable, during the fragmentation of bacterial DNA that accompanies DNA packaging by transducing phages.

10. An auxotroph has a *requirement* not characteristic of the wildtype (prototrophic) strain. A utilization mutant does not have such a requirement but is more or less restricted in *optional* sources of carbon, nitrogen, phosphate, and the like.

11. In transformation, fragments of bacterial DNA must become part of the recipient's chromosome in order to become permanent—that is, to replicate. Plasmids, on the other hand, can replicate without integration, making transformation more efficient once they enter cells.

12. Transposable elements of bacteria cannot themselves leave one cell and enter another; they are transferred between two locations (on DNA) within the same cell. They never have an independent existence, but are always part of another DNA molecule. Their mobility depends upon a genetic organization and a transposase enzyme quite different from standard phages and plasmids. Some phages, however, share some features of transposable elements as they integrate into the bacterial chromosome and excise from it.

13. The bacterial sexual transfer involves only a conjugation tube, while yeast gametes fuse in their entirety. Bacteria pass only *part* of their DNA to the recipient cell in a polarized fashion. The fusion of yeast gametes assures that all genes of both gametes are merged in the zygote. Finally, recombination of the donor and recipient DNA of bacteria is nonreciprocal, since only one product of the recombination event survives; the excess DNA is degraded. In yeast, meiosis conserves all the genetic material, and recombination events are reciprocal, in which both products of each chiasma survive among the meiotic products.

CHAPTER 18
ANSWERS TO PROBLEMS

1. None of the streptomycin-resistant transformants could use galactose for growth. Therefore, few or no fragments of DNA from the donor carry both the determinant for streptomycin resistance and the determinant for galactose utilization. Independent acquisition of two fragments, each carrying one of the two determinants, is not expected in these samples, because the product of the two transformation frequencies (0.95% for streptomycin and 0.26% for galactose) is too low. By the same reasoning, the percentage of cotransformation for streptomycin resistance and mannitol utilization (about 10% in each case) is much higher than the product of their separate transformation frequency ($0.0095 \times 0.008 = 7.6 \times 10^{-5}$). We conclude that the genes for streptomycin resistance and mannitol utilization are linked on the *S. pneumoniae* chromosome and that the linkage of the genes results in their entry and integration as a unit 10% of the time.

2. In the first plating, which selects for *his*$^+$, about half (54%) of the colonies are *trp*$^+$ and half remain *trp*$^-$, indicating that all the selected (*his*$^+$) F$^-$ exconjugants had acquired the *trp*$^+$ marker during chromosome transfer (the efficiency of integration of a marker, once transferred, is about 50%). This means that the *trp* marker *precedes* the *his* marker during transfer. This is confirmed by selecting for *trp*$^+$ in the second plating. Here, only 14% (10/70) of the selected F$^-$ exconjugants also carry the *his*$^+$ allele, indicating that not all conjugants remained together between the times the *trp* and *his* markers entered the F$^-$ cells. Therefore, the answer is (d).

3. Of the four markers, *b* and *c* must be farthest apart, because they are the only pair that is not cotransduced. Marker *a* is closer to *b* than *d* is (29% vs. 1% cotransduction). In confirmation, marker *d* is closer to *c* than *a* is (50% vs. 2% cotransduction). The order is therefore $b - a - d - c$.

4. The order of mutations is established by noting which two yield the greatest number of wild-type recombinants. These are *b* and *d*. Among the crosses including *b, a,* and *c* only, the "end" mutations are those yielding the greatest number of recombinants—namely, *b* and *c*. Thus the order of these three mutations is $b - a - c$. Marker *d* must follow *c*, at the opposite end from *b* by the first consideration above. The map distances are $b - 0.2 - a - 0.1 - c - 0.3 - d$. The figures are double the percentages of wild-type recombinants, because wild types represent only the selected half of the recombinants: double mutants are selected against. The map order may be given as $b - a - c - d$ or $d - c - a - b$. Note that the map distances for larger distances are not additive: the cross $a \times d$ yields a distance of 0.5 units (2×0.25), where the sum of the smaller distances is 0.4 map units. This is common in phage crosses, leading at times to errors in ordering mutations by this method.

5. $- r228 - 0.14 - r145 - 0.10 - (r295, r168) -$

    Note that (i) the problem gave the percentage recombination (= map units), not the percentage of wild-type recombinants; and (ii) the mutations *r295* and *r168* are 0.013 map units apart but cannot be ordered with respect to the other markers with the data given.

6. *r:* V; *s:* I; *t:* IV; *u:* III

7. *dhl,* because it is cotransduced with *aceF* more frequently than *leu* is.

8. *pps* and *man* are far apart; *aroD* is close to *pps;* and *gurA* is close to *man* inside the interval defined by *pps* and *man*.

9. At 12 minutes, 9 minutes after *c,* and 2 minutes before *a.*

10. The prophage DNA, when it enters the F^- recipient, encounters an environment free of phage repressor. Induction of the prophage, followed by lysis of the F^- cell ensues, guaranteeing the death of all potential recombinants. (This phenomenon is called **zygotic induction** of the phage.)

11. Below, the correct order is shown, with donor and recipient arranged to display the position of recombination events. In all cases, the first event picks up thr^+, and the other events yield the genotypes as given.

Donor:	thr^+		$leuY^-$		$leuX^+$	
		1	2	3	4	
Recipient:	thr^-		$leuY^+$		$leuX^-$	

    4 thr^+ $leuY^+$ $leuX^+$ Recombination at $1 + 2 + 3 + 4$

    30 thr^+ $leuY^+$ $leuX^-$ Recombination at $1 + 2$

    20 thr^+ $leuY^-$ $leuX^+$ Recombination at $1 + 4$

    16 thr^+ $leuY^-$ $leuX^-$ Recombination at $1 + 3$

12. The overlaps indicated by the 0's lead to the map below, in which the deletion letters are given above the deletion line. All deletions must be unitary (not broken) segments:

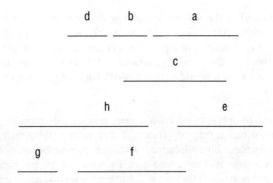

CHAPTER 19
ANSWERS TO QUESTIONS

1. Operator: a *cis*-acting segment of a gene lying 5′ to the transcribed region that, with a repressor protein, controls the ability of RNA polymerase to transcribe the gene. Attenuator: a segment of a gene lying between the start of transcription and the first coding region that controls, in response to environmental conditions, the continuation of transcription into the coding regions. Both of these terms apply mainly to prokaryotic systems. Enhancer: a segment of DNA near a gene in eukaryotes, on the 3′ or the 5′ side, that when bound to an activator protein facilitates binding of RNA polymerase to the promoter. Because it may act at a considerable distance from its target gene, the DNA must be bent or looped to bring the proteins bound to the enhancer and the promoter together.

2. A repressor protein blocks transcription in response to environmental conditions; an activator promotes or facilitates transcription.

3. An operator is a *cis*-acting site, a target in the DNA for a repressor, a negatively acting protein that works in *trans*.

4. Loss of the ability to bind the CAP protein would render haploid cells uninducible for β-galactosidase under any circumstances, because CAP binding is required for transcription. Merodiploids would reveal that this mutation was *cis*-dominant.

5. Two kinds of mutations fit this description: *lacP* mutations, and mutations in the CAP binding site. Both lie in *cis*-acting sequences. The first binds RNA polymerase; the second binds the positively-acting CAP. Both interactions must prevail if transcription is to take place.

6. F′*lac* factors, having genes on the plasmid that duplicate the *lac* region of the chromosome, allow one to introduce a second copy of the *lac* genes into a single cell. If one observes the phenotype of a partial diploid, one may determine dominance relationships. The most important outcome of merodiploid studies was the distinction between dominance and *cis*-dominance—one that depends upon the arrangement of mutations, rather than simply on the presence of the mutations involved.

7. Glucose is readily used as a carbon source by *E. coli,* and, in the presence of glucose bacteria have very low levels of cAMP. Without cAMP, the binding of the CAP protein to the DNA of the operon, needed to activate transcription of the *lac* operon, cannot occur. Therefore, no β-galactosidase is made, even if lactose is present. The system assures that good carbon sources are exhausted before the cell expends energy in making new enzymes, such as β-galactosidase.

8. As indicated in the answer to question 2, controlling proteins may block or activate transcription. Both mechanisms commonly act on single genes or operons; the *lac* operon is an example, in which the repressor protein, the product of *lacI*, represses in the absence of the inducing metabolite (allolactose), and the CAP protein activates transcription when bound to cyclic AMP; both are essential for normal regulation of *lac* proteins.

9. Mutations of operators are *cis*-dominant because they must be physically attached to the transcribed region that they control if they are to display their effects. Forms of the repressor unable to bind allolactose, without impairing the ability of the repressor to bind to the *lac* operator, will be at least partially dominant. Such a mutant protein would repress the *lac* genes to which they are attached even if normal repressor were present.

10. An attenuator sequence, in the transcribed region of DNA just before the protein-coding sequence, interrupts mRNA formation under conditions in which gene expression is inappropriate. The transcript made from this sequence does so by adopting, in the appropriate conditions, a folded structure that stops the RNA polymerase ahead of it from continuing transcription. The formation of the structure is indirectly controlled by ribosomal activity behind the RNA polymerase.

11. Control of gene expression can be exerted in a number of ways after transcription. One example is the control of the translation of an mRNA by sequences in the 5′ mRNA leader, which might block ribosome progress down the mRNA in a controlled fashion. These impediments may be an upstream coding sequence that precludes reading of the main coding sequence downstream or the adoption of a folded, base-paired structure that limits access of the ribosomes to the 5′ end of the mRNA. Beyond these, control of the stability of the mRNA or the protein encoded by it are common mechanisms that control the expression of the gene. Finally, feedback (allosteric) inhibition works strictly at the level of enzyme action, and is usually the most sensitive controlling mechanism of a metabolic pathway.

12. (i) One mutation that might cause the constitutive phenotype is a mutation in an operator of the enzyme-determining gene that makes it insensitive to a repressor. The second is a mutation impairing the structure of the repressor. Such mutations can range from deletion of the repressor gene to subtler mutations impairing binding of the repressor to the operator. (ii) A mutation in an arginine biosynthetic enzyme, not sufficient to cause a requirement for arginine, but enough to reduce the amount of arginine in the cell, could thereby activate a regulatory response by a normal regulatory system.

CHAPTER 20
ANSWERS TO REVIEW QUESTIONS

1. Overhangs result from staggered cuts by restriction enzymes. In making recombinant molecules, the overhangs promote ligation of DNAs having similar ends, exclude ligation with DNA having different overhangs (characteristic of other restriction enzymes), and minimize ligation with blunt ends. The overhangs therefore yield highly significant specificity in making recombinant molecules.

2. Three maps are compatible with the information; they differ in which *Sma*I fragment is in the middle. Of the three pieces of the *Sma*I digest, two (2.5, 2.0 kb) persist in the double digest; these cannot have a *Pst*I site. Because *Pst*I yields only two pieces, its site must be 1.0 kb away from a *Sma*I site within the 4.5 kb *Sma*I piece, which it cuts into 3.5 and 1.0 kb fragments. End-labelling of the original piece of DNA, and then digesting it with *Sma*I, would tell us directly which *Sma*I fragments were terminal and decide the issue.

3. Because the cell methylates and thereby protects the recognition sites of its own restriction enzymes.

4. One would make an antibody to the protein and a cDNA library from the mRNA of the liver. It is likely that the mRNA is abundant, given the abundance of the protein. Colonies of bacteria that are lysogenic for a phage λ vector carrying the cDNA library are induced to lyse; the plaques are then probed with the antibody and labelled protein A to recognize plaques containing the protein or fragment thereof. The phage can be propagated from the original plate and the cDNA retrieved from the phage DNA.

5. With a shuttle cloning vector able to replicate in both yeast and *E. coli,* one should make a library of the genomic DNA of wild-type yeast. The library can then be applied to the His⁻ yeast, with transformants plated on minimal medium to select His⁺ cells. The colonies are then isolated, grown, and the plasmid retrieved from them. They should have the DNA of interest; however, one must check carefully to be sure that another gene has not rendered the mutant cells prototrophic.

6. The scientist should think about making a DNA library of the obscure bacterium in a plasmid that replicates well in *E. coli.* This library, replicating in *E. coli,* should then be screened in some way for colonies that express the protein. This is probably workable, given the fact that the protein of interest is from a prokaryote. The screening might be with an antibody to the protein. However, the scientist might have some way of selecting for the activity of the protein by, for instance, asking the bacteria to use the toxic material as a source of carbon or energy or nitrogen. This would work if the plasmid were to have a strong promoter (for high expression) adjacent to the site in which the library was cloned. Alternatively, the toxin itself might be a selective agent if *E. coli* without the gene were sensitive to it.

7. A useful plasmid vector must have a marker for selection in *E. coli,* a multiple cloning site, and a replication origin that assures propagation as a plasmid in *E. coli.*

8. An 8-base recognition site will appear, on the average, once in about 65.5 kb (4^8). *E. coli* DNA, about 4200 kb per genome, would be cut into about 64 fragments, although they would range greatly in size.

Glossary

activator   In regulatory systems, a protein required for the transcription of a gene. *See trans*-acting factor.

active site   The point on the surface of an enzyme that binds substrates and catalyzes their transformation to products.

additive genetic variance   Genetic variance resulting from the average effects of different alleles in a population.

alkylating agents   A class of chemical mutagens that add methyl and ethyl groups to nitrogenous bases.

alleles   Alternative forms of a gene, such as wild-type and various mutations.

allolactose   The small-molecule inducer, derived from lactose, that actually binds to the *lac* repressor and renders it unable to bind to the *lacO* site of the *lac* operon.

allozyme   An enzyme variant encoded by a particular allele.

alpha carbon   The carbon atom to which both the amino and carboxyl groups of amino acids are attached.

alternation of generations   A term describing a haplo-diploid life cycle, in which both haploid and diploid vegetative forms are found.

amino acid   The repeating unit of polypeptides. There are 20 different kinds.

amino group   In amino acids, the -NH_2 group attached to the alpha carbon.

aminoacyl-tRNA   A tRNA molecule to which a cognate amino acid is attached. The tRNA portion, with its anticodon, serves as an adaptor for the amino acid in reading the codons of mRNA.

aminoacyl-tRNA synthetases   The class of enzymes that covalently joins a tRNA with the cognate amino acid; that is, the amino acid appropriate to the tRNA anticodon.

amniocentesis   A method by which amniotic fluid, surrounding a fetus, is withdrawn by a needle. Cells in this fluid allow diagnosis of chromosomal aberrations and other genetic abnormalities.

amplify   To cause replication to yield many copies, as of a clone of recombinant DNA in bacteria harboring a recombinant plasmid

anaphase   The stage of cell division at which daughter chromosomes (mitosis or Meiosis II) or homologs (Meiosis I) are separated and move to the poles of the spindle.

aneuploid   An organism or cell with more or fewer chromosomes than normal, but in numbers less than a complete set. Example: trisomic, with 2N + 1 chromosomes.

anticodon   The nucleotide triplet in an amino-acyl tRNA molecule that pairs with a corresponding codon in mRNA during translation, thereby locating the amino acid properly in the growing polypeptide.

antigen   Any protein that provokes the appearance of an antibody in a suitable animal (e.g., rabbit, mouse) when injected into it.

ascus, ascospore   In ascomycetous fungi such as yeast and *Neurospora,* the biological form of a tetrad (ascus) and meiotic products (ascospores).

***att* site**   In the *trp* system of *E. coli,* a portion of the mRNA leader (and the DNA encoding the leader) in which an upstream open reading frame regulates the continuation of transcription of an mRNA molecule according to the tryptophan status of the cell.

attenuation   A regulatory mechanism involving control of translation of an upstream open reading frame in an mRNA leader by the availability of an amino acid. This affects the pairing behavior of leader segments, which in turn controls the probability of termination of the transcript before RNA polymerase reaches structural gene information in the operon. *See att* site.

autosome   Any chromosome that is not a sex chromosome.

auxotroph   A mutant requiring a nutritional supplement not needed by the wild-type strain. *See* prototroph.

bacteriophage   A bacterial virus.

Barr body   A condensed, inactive X chromosome found in interphase nuclei of XX individuals of mammalian species.

base analog   A nitrogenous base chemically similar to the normal bases that can be incorporated into DNA or that can interfere with nucleotide or DNA synthesis owing to its resemblance to a normal base.

base-pair substitution   A class of mutation in which one base pair is replaced by another base pair. Includes transitions and transversions (q.v.).

binding site   A term used in regulatory studies to designate a position on a DNA molecule that specifically binds a *trans*-acting protein. (Used in many other contexts in descriptions of macromolecules.)

biparental inheritance   Inheritance in the context of sexual reproduction, implying the contribution of two parents to individual offspring.

bivalent   A configuration in meiosis in which two homologs, each represented by two sister chromatids, are paired to form a four-stranded structure.

blunt cut   An endonuclease scission of a DNA duplex at the same point in both strands, leaving no 3' or 5' overhang.

broad-sense heritability   The proportion of variation having a genetic basis, including additive effects of alleles, dominance effects, and interactions among genes.

5-bromouracil   A chemical analogue of thymine, often used as a base-substitution mutagen.

C-terminal   When speaking of polypeptides, the end with a free carboxyl group; the amino acid added last in translation.

CAP   Catabolite activator protein of *E. coli,* a *trans*-activating factor that, when it binds cyclic AMP, activates genes encoding enzymes required for the utilization of poor carbon sources.

cap   7-methylguanosine phosphate, added at the 5' end of mRNA of eukaryotes.

carboxyl group   In amino acids, the -COOH group attached to the alpha carbon.

carrier   A heterozygote bearing a recessive gene.

cascade (regulatory)   A regulatory system in which several proteins interact sequentially in transmitting an environmental signal to the gene whose transcription is to be regulated.

catalyst   An agent that accelerates a chemical reaction without being consumed by it. Enzymes are biological catalysts.

cDNA   A DNA made as a copy of an RNA (usually mRNA) by the enzyme reverse transcriptase.

cell cycle   The sequence of periods comprising G_1, S (DNA synthesis), G_2 and M (mitosis) required for a new cell to produce two new cells in turn.

centimorgan   One map unit (1% recombination) on a genetic map.

central dogma   The rule that information flows only from DNA to RNA, and from RNA to protein. Relaxed more recently with the demonstration that certain viruses can transcribe RNA into DNA.

centriole   A small organelle found at the poles of the meiotic and meiotic spindle which organizes the spindle fibers. The centriole and the surrounding area are known as the centrosome.

centromere   The constricted region of a eukaryotic chromosome to which spindle fibers attach during mitosis or meiosis. The centromere is required for proper metaphase positioning and anaphase movement.

centrosome   *See* centriole.

chiasma   The cytologically visible point of crossing over seen in late Prophase I of meiosis. Signifies that recombination has taken place between homologous, non-sister chromatids of a bivalent.

chiasma interference   (Usually referred to simply as interference) The inhibitory effect of one crossover event upon the occurrence of another nearby in the same bivalent. It does not pertain to strand-choice. *See* chromatid interference.

chi-squared test   A statistical test designed to compare the expected and observed numbers of events or individuals in different categories in testing a hypothesis, and to yield a probability value (P) that permits judgement of the deviations as significant or insignificant. Includes the use of degrees of freedom, which in simple cases is one less than the number of categories.

chromatid   One of the two sister products of chromosomal replication, before the centromeres visibly divide and the sister chromatids (strands) separate.

chromatid interference   (Rarely observed) The negative effect of the chromatids used in one crossover upon the use of either of them in a second crossover in the same bivalent. *See* chiasma interference.

chromatin   The substance of eukaryotic chromosomes, including DNA, histones, nonhistone proteins and a small amount of RNA.

chromosome   The linear, gene-bearing units of eukaryotes, localized in the nucleus. Applied informally to the indispensable circular DNA molecule of prokaryotes and the DNA of bacteriophage and viruses. The term is not applied to plasmids.

chromosome interference   Chiasma interference (q.v.)

cis   A prefix meaning "on the same side." In gene arrangements, mutants that are *cis* to one another are on the same homolog of a pair. Synonym: **coupling.**

***cis*-acting site**   A site attached physically to a particular DNA sequence that is required for the binding of a *trans*-acting regulatory protein.

***cis*-dominance**   The ability of a *cis*-acting site to exert control on transcription of downstream DNA regardless of a different allele of the *cis*-acting site at another location in the genome or cell.

***cis-trans* test**   A complementation test. Determines allelism or non-allelism of two mutations, assuming that they are recessive. The criterion of allelism is the inability of two mutations, in *trans*, to restore the wild-type phenotype.

cistron   A gene, defined functionally by complementation. A segment of a chromosome within which two (allelic) mutations do not complement; a region coding for one polypeptide.

clone   A group of identical organisms derived by mitosis or by asexual propagation. Also, a group of identical DNA molecules arising through replication of a plasmid or other small DNA.

coadaptation   The functional adaptation of one gene or a group of genes to one another by a coevolutionary process.

codon   A nucleotide triplet specifying an amino acid or a nonsense triplet.

coding ratio   The number of nucleotides in sequence required to specify amino acids. In a triplet code, the coding ratio is three.

codominance   A phenotypic relationship between alleles in which both can be detected as qualitatively distinct. Example: blood-group antigens.

coefficient of coincidence   The ratio of the observed frequency of double crossover progeny to that predicted from multiplying the separate recombination probabilities of the two chromosome regions involved.

coenocyte   A tissue or organism having more than one nucleus per cell; a syncytium. A common condition in filamentous fungi. *See* heterokaryon.

cognate   In informational terms, "corresponding to." The amino acid histidine is cognate to tRNA^{his}.

cointegrate   Two circular DNA molecules fused into a single circle. Example: the Hfr chromosome of *E. coli*.

colchicine   A drug that prevents the formation of the mitotic or meiotic spindles.

colinearity   Point-for-point correspondence of two linear sequences, such as gene and protein.

combinatorial   A term referring to different arrays of a few variable elements, such as the different sets of *cis*-acting sites of different operons.

competence   In transformation, the state of cells able to take up exogenous DNA.

complementary   In pairs, having forms or information in mirrored or cast/mold correspondence. Complementary strands of DNA have defined base pairs at each point. Complementary pairs of recombinants have a mutant/wildtype correspondence for each allele. Complementary, contacting surfaces of polypeptides conform to one another in a cast/mold fashion.

complementation   The ability of two genetic elements (chromosomes or nuclei), each carrying a single mutation, to confer a wild-type phenotype when they are present in the same cell. Used also to describe the ability of a recombinant DNA, when introduced into a mutant cell, to impart a normal phenotype. *See* dominance.

conditional mutation   A mutation whose expression depends upon special conditions of growth. Example: temperature-conditional mutations, imparting a mutant phenotype only at an elevated temperature.

conidium (-a)   The asexual spore of *Neurospora* and other filamentous fungi.

conjugation   In bacteria, joining of cells via a cytoplasmic bridge, followed by transfer of DNA or a plasmid from donor to recipient.

consensus site   A short nucleotide sequence in multiple copies in a gene or segment of DNA, or a sequence shared by several genes that implies that they are controlled by common *trans*-acting factors.

conserved region   A region of a protein or a nucleotide sequence that remains similar over evolutionary time.

constitutive   Always present. Commonly applied to enzymes that cannot be repressed owing to mutation of a regulatory gene or *cis*-acting sequence.

continuous traits   Traits showing no discrete categories.

cos site   A site in the DNA of phage lambda which suffers a staggered cut (12 base-pairs apart on the two strands) prior to packaging the DNA into phage heads.

cotransduction   Simultaneous transduction of two bacterial genes, implying that they are borne by the same transducing phage particle.

cotransformation   Simultaneous transformation and integration of two or more genetic markers.

counterselection   Selection against an undesired phenotype, as in selection against parental cells in conjugation or transduction experiments.

coupling   The *cis* arrangement of two markers on the same homolog or other genetic element.

cross-feeding   The use of a metabolite accumulated and excreted by one type of cell by another cell which can use it for growth or further metabolism.

crossing over   The exact, reciprocal exchange of homologous chromosome parts at chiasmata in Prophase I of meiosis. Extended to denote any normal (usually homologous) recombination in bacteriophage and bacteria.

cyclic AMP (cAMP)   A small molecule, derived from ATP, with a phosphate bonded to both the 2′ and 3′ carbons of the ribose. The concentration of cAMP is indicative of the carbon status of prokaryotic cells.

cytoplasmic inheritance   A pattern of inheritance reflecting a replicating entity separate from the chromosomes, usually the DNA of a symbiont or an organelle.

daughter-strand gap repair   *See* recombination repair.

degrees of freedom   The number of variables, or categories, free to vary independently.

deletion (deficiency)   A loss of a segment of DNA, or a DNA which has suffered such a

loss. Usually applied to deletions larger than 4 nucleotides, the latter being frameshifts.

derepression   Relief of negative control by removal of a repressor molecule. Applied to a physiological response to an effector, or to genetic change leading to unresponsiveness to negative control.

diakinesis   The last stage of prophase I of meiosis, at which bivalents achieve their greatest condensation prior to lining up on the metaphase plate.

diploid   A cell with two sets of chromosomes (2N), or the 2N condition.

diplotene   The stage of Prophase I of meiosis following Pachytene, during which the synaptonemal complex disappears, and chromatids separate except at chiasmata.

direct repeat   A DNA sequence found in the same orientation at two different locations on a DNA molecule. These are often found as tandem duplications, one immediately next to the other in a head-to-tail arrangement.

disomy   The condition of a cell, normally haploid, with one extra chromosome (1N + 1).

distal   Applied to the location of a gene, away from the centromere. *See* proximal.

disulfide   Two sulfur atoms in covalent linkage. In proteins, two S-containing cysteines in different polypeptides may form a disulfide that stablizes the quaternary structure.

DNA damage   Abnormal, chemically altered DNA. Distinct from mutation, which is informationally changed, but chemically normal.

dominance   The phenotypic masking of one allele (the recessive) by another allele (the dominant) in the same cell.

dosage compensation   A class of mechanisms by which the homo- and heterogametic sexes become equivalent in the magnitude of expression of sex-linked genes.

double-strand gap repair model   A model of recombination by which a gap in one DNA initiates recombination in a homologous DNA. Its distinctive character is the replacement of the deleted material by using both DNA chains of the homolog as templates.

Down's syndrome   A human congenital disorder arising from trisomy for chromosome 21.

effector   A small molecule that binds to a regulatory protein and either activates or inactivates its regulatory capability. Example: cAMP is an effector for catabolite repression owing to its effect on CAP.

electrophoresis   A technique in which proteins or DNAs are placed at a point in a semisolid medium (agarose, polyacrylamide, starch gel) and subjected to an electrical potential. The net charge of a protein, determined by the relative numbers of charged R groups, determines its direction and rate of movement through the gel. The technique permits separation of closely related proteins. DNAs move to the positive pole in electrophoresis, which separates linear molecules on the basis of size.

endonuclease   An enzyme that cuts a nucleic acid within the chain. (An exonuclease attacks nucleic acid chains at one or the other end.) *See* restriction enzyme.

endoreduplication   The replication of a chromosome or chromosome set without a mitotic or meiotic distribution of daughter chromosomes. Leads to polyploidy or polytene chromosomes.

enhancer   A segment of DNA that increases expression of a gene to which it is attached (i.e., in *cis*). Can exert its effect in downstream or upstream positions, and in either orientation. In eukaryotic microbes, **upstream activation sequences** are sometimes called enhancers even when they have not been tested for the criteria stated.

environmental variance   Environmentally induced variation.

enzyme   A biological catalyst, usually a protein, that vastly increases the rate of metabolic reactions over the spontaneous rate at ambient temperature.

epistasis   The masking effect or interference of one gene upon the phenotypic expression of a non-allelic gene or mutation in the same genome. *See* dominance.

ethidium bromide   A fluorescent molecule used to stain DNA in gel electophoresis experiments.

euchromatin   Lightly staining, less condensed chromatin suggesting its metabolic activity. *See* heterochromatin.

euploid   The condition of an organism or cell having the basic haploid chromosome number or an exact multiple thereof.

excision repair   Repair of heteroduplex DNA by excision of the damaged strand (if chemically identifiable) and resynthesis of a new one, using the remaining strand as a template.

exconjugant   A female bacterial cell that has received DNA from an F^+ or Hfr cell during conjugation.

exon   A segment of a gene or transcript that is present in the mature mRNA. Applied to genes in which introns, or intervening sequences (q.v.), are found. The term implies an "expressed unit."

expressivity   The intensity of expression of a gene. Applied to genes that have variable expression in similar or true-breeding stocks. *See* penetrance.

F factor   A 94 kb DNA, existing as a plasmid or as a segment of the bacterial chromosome, that confers the ability of the host to conjugate with other cells. The F factor promotes its own transfer to other cells if it is in its plasmid form (autonomous), or the transfer of bacterial genes to which it is attached if it is integrated into the chromosome (as in Hfr bacteria). F^+ bacteria have an autonomous F factor; F^- bacteria have no F DNA; **Hfr** bacteria have an integrated F factor; and F' bacteria have an autonomous F factor that carries some bacterial genes. The term F' is applied to both the bacterium and to the modified F factor itself.

F_1, F_2 generations   First and second filial generations, usually starting with true-breeding diploid parents. Inter-crossing or self-crossing the F_1 progeny generates the F_2 generation.

feedback inhibition   The inhibition of an early enzyme of a metabolic pathway by the small-molecule endproduct of the pathway.

fertilization   The fusion of haploid gametes and their nuclei to form a diploid zygote.

first-division segregation   Segregation of a pair of alleles at the first meiotic division. *See* second-division segregation.

fitness   The number and quality of offspring a genotype produces, relative to other individuals in the population. Measured by the differential contribution of genes to the next generation.

frameshift mutation   Insertion or deletion of 1 to 4 bases in a DNA coding sequence, leading, in the case of all but $+3$ or -3, to an alteration of the normal reading frame.

G_1, G_2   Periods ("gaps") of the cell cycle between mitosis and S-phase, and between S-phase and mitosis, respectively. The G_1 period is usually considerably longer than the G_2 phase

gamete   A haploid cell specialized to fuse with another haploid cell (fertilization) in a sexual life cycle.

gene   A segment of DNA encoding a polypeptide or one of the RNAs involved in translation. Includes the 5' and 3' sequences required for control, transcription, mRNA processing, etc. *See* cistron.

gene arrangement   A term describing the location of alleles of non-allelic genes on two homologs of a diploid or heterokaryotic cell. With respect to two genes, the arrangements are *cis* or *trans* (coupling or repulsion).

gene conversion   A rarely occurring conversion of one allele of a heterozygous bivalent to the other allele. Takes place by correction of heteroduplexes formed between non-sister chromatids or by double-strand gap repair.

gene imbalance   Abnormal ratio of genes in aneuploids or organisms with chromosomal aberrations.

gene order   The order of genes on a chromosome, represented by a genetic map.

gene pool   All of the alleles present in a population. No assumption is made about the distribution of alleles in individuals.

genetic drift   Random processes in populations that lead to changes in allele frequencies.

genetic engineering   The field that relies on forming recombinant DNA molecules in the test tube, and cloning them in bacteria.

genome   One complete set of genes as seen in phage, bacteria, or the haploid phase of eukaryotes. Often used to designate the full genetic complement of diploid organisms.

genotype   The genetic constitution of a cell or organism, often designated only by alleles of genes of interest.

genotype-environment interactions   Seen when genotypes respond differently to changes in an environmental variable.

genotypic variance   Genetically induced variation.

germ cell   The diploid cells of diploid organisms (e.g., spermatocytes, oocytes) that will, via the meiotic process, form gametes.

global control   A term describing coordinate regulation of large repertories of genes in response to major environmental influences, such as carbon starvation, UV damage (SOS repair), and heat shock.

haploid   A cell or organism with a single set of genes; 1N.

Hardy-Weinberg law   Alleles in a population are distributed to individuals according to the the sum of the frequencies squared [$(p + q)^2$]. Allele and genotype equilibria are reached in a single generation.

hemizygous   Applied to diploid organisms with a single representative of a given gene. Example: X-borne genes of (heterogametic) males.

hemophilia   A genetic deficiency in blood clotting.

heterochromatin   Densely staining, condensed, and inactive chromatin. *See* euchromatin.

heteroduplex   A DNA with a mismatch between the two strands, or with damage on one strand. Includes not only base-pair mismatches, but deletions and insertions.

heterogametic   A term describing the sex having different sex chromosomes.

heterokaryon   A cell with genetically different, distinct nuclei. Common in coenocytic fungi.

heterozygote   A diploid cell or organism having different alleles of a given gene.

heterozygote superiority   Heterozygotes have higher fitness than either homozygote. This results in an equilibrium of the alleles, although selection constantly occurs against homozygotes each generation.

Hfr   A male bacterium with an integrated F factor, leading to <u>h</u>igh <u>f</u>requency of <u>r</u>ecombination in crosses with F⁻ cells.

histones   Basic proteins complexed, as octamers, to the DNA of eukaryotic organisms.

Holliday structure   An intermediate in recombination in which one nucleotide chain of each of two homologous DNAs are interchanged at one point, leading to a "half-chiasma." Must be resolved before segregation or replication.

homogametic   A term describing the sex having two sex chromosomes of the same type, such as XX females.

homokaryon   A coenocyte in which the nuclei are genetically identical.

homologous chromosomes (homologs) Chromosomes that carry the same information except for allelic differences. In diploids, one homolog is contributed by the maternal parent, the other by the paternal parent.

homozygote   A cell or organism (diploid) having identical alleles of a given gene.

hybrid   The offspring of genetically different parents.

hybridize   In molecular biology, the pairing of complementary strands of homolgous DNA molecules, as in the probing of a Southern blot.

hydrogen bond   A hydrogen atom shared between two distinct chemical groups, leading to a weak bond. Found between adenine and thymine (2 H-bonds) and between guanine and cytosine (3 H-bonds) in DNA.

hypha (-ae)   The cellular filament characteristic of filamentous fungi.

inactive X   The X chromosome of mammalian females that becomes a condensed

Barr body, and that is metabolically inert, starting in early embryogenesis.

inbreeding coefficient   Denoted by F, the inbreeding coefficient, $(2pq - H_1)/2pq$, indicates the intensity of inbreeding in a population. When observed (H_1) and expected ($2pq$) frequencies of heterozygotes are identical, F equals zero (no inbreeding). As the observed number of heterozygotes approaches 0, F approaches 1.

incomplete dominance   Semidominance; the incomplete phenotypic masking of one allele by another, applied to characters in which the alleles achieve a quantitative compromise in expression.

independent assortment   Independent segregation of two or more pairs of alleles, leading to equal numbers of parental and recombinant meiotic products. Independent assortment is assured by independent orientation of different bivalents during Metaphase I.

indirect selection   Artificial selection in which variants are selected not from the selective environment, but the from the untreated population on the basis of the behavior of the sample that is exposed. Replica plating, for instance, affords this opportunity.

inducer   A small molecule that evokes gene expression, either by blocking a repressor action or by activating a *trans*-acting factor.

infective conversion   Conjugal transfer of the F factor from F^+ to F^- bacteria, converting them to F^+.

informational macromolecule   Macromolecules with a specified sequence of variable repeating units. Usually applied to proteins and nucleic acids.

initiation factors   In protein synthesis, the proteins acting on mRNA, ribosomes, and the initiating amino-acyl tRNA to assemble a complex that will begin translating the mRNA.

insert   A piece of DNA introduced into a vector for cloning or amplification.

insertion sequence (IS)   A short piece of DNA 800–1500 base-pairs long with inverted terminal repeats that can transpose from one location to another in the genome. The smallest and simplest of the transposable elements.

integrase   An enzyme capable of inserting phage DNA into a chromosome.

intercalating agent   A planar organic molecule that can insert itself between base-pairs of DNA. This causes distortion of DNA during replication or repair and induces frameshifts in the new DNA.

interference   *See* chiasma interference and chromosome interference.

intergenic complementation   Complementation (q.v.). *See* intragenic complementation.

interphase   The period of the cell cycle (G_1, S, and G_2) between the end of one mitotic phase and the beginning of the next. In interphase, chromosomes are not distinctly visible.

intragenic complementation   Results from the interaction of allelic polypeptides in a cell having the corresponding mutant alleles in *trans*. The allelic polypeptides mutually stabilize or activate one another in a mixed multimer such that the function of the multimer is partially restored.

intron   A segment of DNA in the transcribed region of a gene that is removed from the transcript immediately after an mRNA molecule is made. Originally, but rarely now, called an "<u>in</u>tervening sequence," from which the term was derived.

inversion   A DNA with a segment in the reverse order, compared to a standard sequence.

inverted repeat   A nucleotide sequence found in identical or almost identical form, but inverted, such as those at opposite ends of transposons.

isomerization   Literally, the shift between two defined states of a molecule. Loosely used to describe the continuous change of the Holliday structure between two extreme forms, each with a different pair of crossed nucleotide chains.

karyotype   The characteristic appearance of the chromosomes of a cell spread out at metaphase, at which division is arrested. Used for determining chromosome abnormalities.

kappa particle   A symbiotic bacterium found in *Paramecium* that confers upon the host the ability to secrete a toxin lethal to *Paramecia* lacking the kappa particle.

killer *Paramecium*.   Kappa-bearing *Paramecia*. *See* kappa particle.

kilobase   1000 bases or 1000 base-pairs.

kinetochore   The aggregate of proteins found on centromeric DNA that is required for spindle attachment.

Klinefelter syndrome   A human male phenotype characterized by sterility and disturbances of secondary sex characteristics conferred by the XXY genotype.

***lac* operon**   A continuous ensemble of genes and their *cis*-controlling elements devoted to the utilization of lactose. Includes the gene for β-galactosidase.

lagging strand   In DNA replication, the new nucleotide chain that is made in segments, growing away from the replication fork.

leader   In mRNA, the segment of the nucleotide chain between the 5′ nucleotide and the first nucleotide of the coding region.

leading strand   In DNA replication, the new nucleotide chain that is made continuously, growing into the replication fork.

leptotene   The earliest stage of Prophase I of meiosis, in which chromosomes first appear microscopically.

library   A cloned, heterogeneous set of DNA or cDNA fragments representing the genome or the mRNA population of a given organism. A mini-library is a library made from a lesser fraction of the source material. Libraries are sometimes called banks.

life cycle   The complete sequence of stages required to replicate an organism. The asexual life cycle (either haploid or diploid) goes through various biological forms (growing, dormant, etc.) without change of ploidy. The sexual life cycle requires an alternation of haploid and diploid phases, accomplished by meiosis and fertilization, respectively.

ligase   An enzyme that links two molecules together; applied to DNA ligase, which makes two DNA nucleotide chains of the same polarity continuous by linking them end to end.

linear tetrad   A genetic term describing the long ascus of fungi. In some species (e.g., *Neurospora*) the derivatives of the first meiotic division are found in the two halves (upper and lower) of the ascus, allowing one to judge the first- or second-division segregation of a pair of alleles.

linkage   The association of markers in a cross that yields less than 50% recombinants. Usually indicates location of the markers on the same chromosome.

linkage map   A representation of the order of genes, with intervals between them proportional to the probability of recombination. *See* map unit.

Lyon hypothesis   Named for Mary Lyon, the hypothesis that dosage compensation (q.v.) in mammals is achieved by inactivation of one of the two X chromosomes in females.

lysogen, lysogeny   A lysogen is a bacterium carrying an inducible prophage. Lysogeny is the process by which an infecting, temperate phage DNA inserts into the bacterial chromosome (or becomes a plasmid) to create a lysogen. *See* lytic cycle.

lytic cycle   The series of events following infection by a temperate phage or induction of a prophage that includes DNA replication, the formation of mature phage and lysis of the cell. An alternative path to lysogeny.

M phase   In the cell cycle, mitosis.

macrolesion   Any major heritable alteration in DNA, such as deletion and insertion (of more than 4 bases), inversion, translocation, or loss of a chromosome.

map unit   One percent recombination between two markers in a cross, when all recombinant progeny are accounted for.

maternal effect   The influence of the mother's genotype upon the phenotype of her offspring. Not persistent to the next generation. *See* cytoplasmic inheritance.

maternal inheritance   Cytoplasmic inheritance in cases in which the inherited trait is derived from the mother. *See* cytoplasmic inheritance.

mating types   Strains of a microorganism differing by the allelic state of one or more genetic loci that determine sexual compatibility. *Neurospora* and yeast have two mating types (*A* or *a* and *a* and *α*, respectively). Mating type can be superimposed on sexual differentiation (male, female), which refers to the morphology and functions of parents and/or gametes.

meiosis   A process comprising one DNA replication and two cell divisions by which a diploid yields four haploid products.

meiotic product   One of four haploid products of meiosis.

meristic traits   Traits measured by counting, the values always being discrete.

merodiploid   Partial diploid. Used to designate the state of bacteria, such as an F′ cell, having more than one copy of one or more genes.

messenger RNA (mRNA)   The RNA derived from transcription of a gene encoding a protein. The transcript (often called "primary transcript") may be processed and spliced, and therefore may contain sequences that the mRNA lacks.

metabolic pathway   A series of enzymes and small-molecular-weight intermediates in which a starting material is converted to another metabolite.

metaphase   Stages of mitosis or meiosis at which the chromosomes, organized as pairs of chromatids (mitosis or meiosis II) or as bivalents (Meiosis I) line up at the metaphase plate in preparation for distribution to the poles in the anaphase that follows.

methylation   The state of a nucleotide chain in which methyl groups ($-CH_3$) have been added. Methylation is confined largely, but not exclusively, to cytosine in natural systems.

migration   The exchange of genes among populations, influenced both by the number of individuals exchanged and the magnitude of differences in allele frequency in the source and recipient populations.

minimal medium   A culture medium containing only the ingredients the wild-type strain of an organism needs for growth. *See* auxotroph.

missense mutation   A base substitution causing an amino acid substitution in the protein that the affected gene encodes.

mitosis   Eukaryotic cell division, starting with the cytological appearance of chromosomes in the nucleus and culminating with the separation of daughter cells. Mitosis does not lead to a change in ploidy. Includes nuclear division (karyokinesis) and cell division (cytokinesis).

mixed infection   Used to describe a mating of bacteriophage, which requires that all bacterial host cells be infected by at least one bacteriophage of each "parent" phage.

monocistronic operon   An operon with only one protein-coding region.

monosomy   The aneuploid state of a cell type, normally diploid, with only one homolog of a particular chromosome.

mosaic   A multicellular organism with a mixture of genetically different cells.

multifunctional polypeptide   A long polypeptide having domains (usually separately folded) with different catalytic functions. Many such polypeptides arise through fusion of the coding regions of genes that were originally distinct.

multimeric protein   A protein with more than one polypeptide chain. All chains may be encoded by the same gene (homomultimer), or they may be encoded by different genes (heteromultimer).

multiple alleles   The existence of more than one allele at a given gene. Usually used when several variants are easily distinguishable, such as the ABO blood groups.

multiple cloning site   A segment of a cloning vector having a series of cloning sites unique to the vector, into which foreign DNA can be inserted.

multivalent   A group of more than two homologous chromosomes in meiosis in which, however, pairing takes place two-by-two in all segments. Seen in aneuploids (2N + n) or polyploids.

mutagen   A chemical or physical agent that causes DNA damage that, when misrepaired or not repaired, leads to mutations.

mutant   A cell or organism carrying a mutation in its genotype.

mutation   A unitary, heritable change in the genes of an organism.

mutation frequency, mutation rate   Mutation frequency is a measure of the number of mutants expressed in terms of total cell number. The mutation rate is the rate at which mutations occur, per cell, per generation.

mutational site   The smallest mutational site is a nucleotide pair.

mycelium   The tangle of hyphae of a fungus, usually arising from a single spore or group of spores. A mycelium is to a hypha what a multicellular organism is to a cell.

N-terminal   In a polypeptide, the end with a free amino group. Natural polypeptides begin to form from the N-terminal end.

narrow-sense heritability   The proportion of variation attributable to additive effects of alleles.

negative control   A mechanism in which regulation is exerted by a repressor that blocks gene expression. *See* positive control.

neutral mutation   A mutation without significant selective disadvantage or benefit to the organism, compared to the prevailing allele(s).

nondisjunction   The failure of pairs of homologs to distribute properly, one to each daughter cell during cell division. Leads to aneuploid products.

nonhistone proteins   In the context of chromatin, all proteins that are complexed with DNA except the histones.

non-Mendelian inheritance   The forms of inheritance that involve non-chromosomal, heritable entities such as organelle DNA. It is generally not used to describe maternal effect. *See* cytoplasmic inheritance.

nonparental ditype   Applied to two genes, a tetrad having only the two recombinant genotypes.

nonsense mutation   A base substitution in a coding region that, transcribed into mRNA, represents one of the three termination codons, UAG, UGA, or UAA. These codons prematurely terminate translation of the polypeptide.

nuclear region, nucleoid   These terms describe the DNA as it exists within a bacterial cell. Often the term nucleus is used to describe this region, although it has no nuclear membrane.

nucleic acid A macromolecule (DNA or RNA) made up of a linear, unbranched sequence of nucleotides.

nucleosome An octamer of histone molecules around which a short segment of DNA is wrapped. This is a common, first-level state of DNA condensation in eukaryotes. Nucleosomes form densely on eukaryotic DNA, one approximately every 140 base-pairs.

nucleotide The repeating unit of nucleic acids, of which there are four kinds.

null allele An allele having no phenotypic expression, usually either yielding no protein, or the protein of which is wholly inactive.

nullisomic Usually, the meiotic or mitotic product of nondisjunction that has no representative of a chromosome (e.g., 2N-2 or 1N-1)

oligonucleotide A short RNA or DNA molecule (usually single-stranded), having 3–75 bases. Applied to chemically synthesized chains or natural degradation products of DNA.

oligopeptide A short polypeptide chain (4–30 amino acids).

operator A DNA region at the 5′ end of an operon that acts as a binding site for a negatively acting repressor. May overlap the promoter.

operon Usually refers to a continuous series of bacterial genes that is transcribed into a single, polycistronic mRNA. The operon includes its *cis*-acting regulatory sites. The mRNA is usually translated into multiple polypeptides. The term can be applied to monocistronic units.

ordered tetrad *See* linear tetrad.

organelle A subcellular molecular aggregate or membrane-bound compartment of specialized function. Only a few types of organelles (e.g., mitochondria and chloroplasts) have DNA.

origin of transfer The first point of bacterial DNA to enter the F⁻ cell during a Hfr × F⁻ mating. It is preceded by part of the F DNA.

pachytene The stage of Prophase I of meiosis in which homologous chromosomes are fully paired throughout their length by way of the synaptonemal complex. Crossing over occurs at this stage.

palindrome A DNA sequence in which both strands of the duplex have the same sequence when read 5′ to 3′; i.e., a tandem inverted repeat. Restriction sites are usually palindromic.

parental ditype In reference to two genes, a tetrad in which only the two parental genotypes are found.

pedigree In general, a family tree. A formal diagram that depicts the inheritance of a trait from one generation to the next.

penetrance A measure of the proportion of genetically similar individuals that show any phenotypic manifestation of a gene they have in common. *See* expressivity.

penicillin enrichment A technique of bacterial genetics in which growing cells are killed, thereby enriching the population for mutants that cannot grow.

peptide bond The bond that links sequential amino acids of a polypeptide. Forms between the carboxyl group (COO^-) of one amino acid to the amino group (NH_3^+) of the next.

peptidyl-tRNA A molecule consisting of a tRNA to which a growing (poly)peptide is attached. It is normally bound to a ribosome, and is a transient entity during translation.

petite colonie Yeast forming small colonies, due to slow growth conferred by impairments of mitochondrial function. Arising from mutations in either chromosomal (**segregational petites**) or mitochondrial (**cytoplasmic petites**) DNA.

phage Bacteriophage. Bacterial virus.

phenotype The appearance or behavior of an organism, as distinct from its genetic constitution (genotype).

phenylketonuria (PKU) A genetic deficiency in humans affecting the disposal of the amino acid phenylalanine.

photoreactivation Repair of pyrimidine dimers (induced by UV light) by a light-activated enzyme that breaks the bonds between the pyrimidines *in situ*.

pilus (-i) A small proteinaceous filament on the external wall of Hfr and F⁺ bacteria. In *E. coli*, pili are required for recognition of F⁻ cells by the Hfr and the formation of the conjugation bridge.

plaque A clear zone of bacterial lysis, made by a spreading bacteriophage infection, in an otherwise turbid lawn of bacteria on an agar surface.

plasmid A circular DNA, usually dispensable to the growth of a cell, that replicates independently of the chromosomes of pro- or eukaryotes.

pleiotropy Multiple phenotypic effects of a gene or mutation.

ploidy The number of sets of chromosomes in a cell or organism.

point mutation Highly localized genetic change. Reserved for base-pair substitutions and frameshift mutations.

poly(A) tail   A sequence of 50–200 ribo-adenylic acids added to mRNA molecules of eukaryotes after transcription is complete.

polycistronic   Having multiple coding regions. Usually used to describe operons.

polymerase (DNA, RNA polymerase)   An enzyme that catalyzes the joining of nucleotides to the 3′ end of a growing nucleotide chain.

polymerase chain reaction (PCR)   A means by which a small amount of a segment of DNA may be amplified, using specific primers, in many sequential rounds of DNA synthesis *in vitro*.

polymorphism   The condition of a genetically heterogeneous population in which there are more than a few percent of certain minority types.

polypeptide   The basic unit of proteins, consisting of an unbranched chain of amino acids.

polysaccharide   A macromolecule made up of sugars as the repeating unit.

polysome   A mRNA molecule having multiple ribosomes on it, each engaged in translating the coding sequence.

polytene chromosome   A chromosome consisting of many copies of the DNA molecule, due to a number of replications without separation of daughter molecules.

population   Those individuals of a species that occur in a geographic area and exhange genes with equal probability.

positive control   A form of control in which a gene is regulated by activation by a *trans*-acting factor binding to a *cis*-acting site.

postmeiotic segregation   The appearance of two genetically different daughter cells in the first mitotic division of a meiotic product. Implies that the DNA for the gene in question in the meiotic product was heteroduplex.

premeiotic S phase   The S phase immediately prior to meiosis, but not considered part of meiosis.

primary structure   The sequence of amino acids in a polypeptide. Extended to apply to the sequence of repeating units in any linear (segment) of a macromolecule.

primase   The enzyme (a form of RNA polymerase) required to form the primers for synthesis of DNA of the lagging strand.

primer   The growing (3′ OH) end of a nucleic acid chain. In more restricted context, primers are the short RNA molecules required for the initiation of synthesis of a segment of DNA.

probe   A molecule that will specifically recognize another molecule in a complex mixture, as an oligonucleotide recognizes a homologous sequence in Southern blotting, or an antibody finds the target antigen in a protein blot.

promoter   A 5′, *cis*-acting segment of a structural gene or operon that acts as a binding site for RNA polymerase and *trans*-acting activators nearby.

prophage   The latent form of a bacteriophage in a lysogenic bacterium. Usually inserted into the bacterial chromosome, but some prophages are plasmids.

prophase   The initial phase of mitosis, when chromosomes condense into cytologically visible threads. Also, the initial phase (Prophase I) of meiosis, in which homologs pair and become visible. Prophase II of meiosis is usually omitted.

prototroph   A strain of an organism that can grow on minimal medium. *See* auxotroph.

proximal   In chromosomes, toward (or nearer) the centromere.

pseudodominance   The expression of a recessive mutation in hemizygous form. Usually used in the context of mutations paired in heterozygotes with a deletion that includes the gene in question.

Punnett square   A chart having gametic genotypes of two parents, one corresponding to rows, the other to columns, by which zygotic genotypes can be rigorously predicted at the intersections.

pyrimidine dimer   A dimer of two sequential pyrimidines on a strand of DNA, formed during UV irradiation. *See* thymidine dimer.

quantitative variation   Genetic variation governed by a number of genes, each having a small effect on the phenotype.

quaternary structure   The number of polypeptides ("subunits") of a protein and their mode of aggregation.

R group   The chemical groups by which amino acids differ. (Used in the description of many homologous series of chemicals.)

random coil   A structural term denoting a disordered, heterogeneous structure of macromolecules. Single-stranded DNA falls into a random coil where it does not form self-complementary regions.

reading frame   In a continuous sequence of nucleotides, the arbitrary setting that determines the triplets that are actually read. In a triplet code, there are three reading frames, only one of which is read in any given gene.

reciprocal crosses   Crosses differing in the sex or mating type that contributes a given allele of a gene.

reciprocal recombination Formation of equal numbers of each pair of complementary recombinants in a cross. Applies to individual recombination events, tetrads, or, statistically, to populations of progeny.

reciprocal translocation An aberrant exchange of parts between two non-homologous chromosomes.

recombinant A haploid meiotic product that is different in genotype from either gamete that gave rise to the diploid from which the meiotic product came. More generally, progeny that differ from parents owing to recombination at meiosis.

recombinant DNA A novel DNA sequence constructed in vitro from segments not found in the DNA preparations used to make it.

recombination repair Repair, during replication, of a gap in a new DNA chain caused by prior damage (e.g., pyrimidine dimer) to the template chain. Use is made of the template strand of the newly synthesized sister duplex to replace the gapped, new strand of the other duplex at the replication fork.

regulatory gene A gene that has evolved to control the degree of expression of another gene.

replica plating A technique of printing colonies, lifted from one solid medium, onto another medium, without changing their positions.

replication fork The point on a replicating DNA molecule at which the two strands of the duplex are separating, and where new DNA chains are being synthesized.

repressor A protein that binds to an operator and blocks transcription of the downstream gene or operon.

repulsion The *trans* orientation of two genes on homologous chromosomes. Two non-allelic mutations are in repulsion if they lie on different homologs. Applies to allelic mutations when they affect different mutational sites.

restriction enzyme An endonuclease that cuts DNA at or near specific sequences of nucleotides that it recognizes. The enzymes used in genetic engineering research (Type II) cut at the recognition sequences.

restriction fragment length polymorphism (RFLP) Owing to mutational differences in the existence of restriction sites, the varying lengths of DNA fragments of related organisms (within a species) having genes in common.

restriction map A map of restriction sites on a DNA molecule or plasmid.

reversion Back-mutation of a mutant to a wild-type phenotype. May occur through restoration of the wild-type nucleotide sequence, through a compensatory, second-site mutation within the same gene, or through mutation of a suppressor gene at another genomic location.

ribosomal RNA A specific class of RNA (3–4 types) found as part of the structure of ribosomes. rRNA is not translated.

ribosome An aggregate of specific RNAs (rRNAs) and proteins that serves as the surface on which translation occurs.

RNA processing The processes by which a primary transcript of a gene has its introns removed (splicing), its m^7G added to the 5′ end (capping) and a poly(A) tail added.

rolling circle Used to describe the sigma mode of DNA replication, in which one chain of a circular DNA is broken and one broken end is peeled away from the other, intact circular DNA chain. The replication fork formed is the site of synthesis of new DNA chains, which requires that the circle rotate ("roll") in relation to the replication fork.

S phase The period of DNA synthesis in the eukaryotic cell cycle.

satellite DNA DNA of which there are many copies, of different base composition in the genome. Seen in higher eukaryotes. The term satellite derives from the separation of this DNA from the "single-copy" DNA after extraction and centrifugation.

second-division segregation Separation of a pair of alleles at anaphase II of meiosis. Occurs because of crossing over between the gene and its centromere prior to Metaphase I.

secondary structure In proteins, the local folding of a polypeptide owing to the formation of H-bonds between atoms of the peptide backbone.

segregation Separation of alleles of a single gene at Anaphase I or II. Leads to 1:1 ratios of alleles among meiotic products.

segregational petite *See* petite colonie.

selection Differential genetic contribution of genotypes to the next generation.

selective coefficient A measure of the disadvantage of one allele compared to another in terms of fitness in a population.

self-fertilization Matings that involve male and female gametes produced by the same individual. This type of mating is the most severe form of inbreeding.

self-incompatibility A mating system in hermaphroditic plants that prevents self-fertilizations and matings among individuals having the same incompatibility type.

semiconservative   Applied to DNA replication, describes the intactness of each original nucleotide chain (half-duplex) in the daughter molecule.

semidominance   *See* incomplete dominance.

sex chromosome   A chromosome of diploids by which the sexes differ in number (X chromosome) and type (Y chromosome), and that participates in sex determination. Often understood to mean the X chromosome.

sex linkage   The inheritance pattern of genes on the X chromosome.

shotgun cloning   Cloning all heterogeneous fragments of DNA that are products of a restriction enzyme digestion.

shuttle vector   A plasmid that will replicate and can be selected for in two kinds of organism, *e.g.*, yeast and *E. coli*. Developed most fully for yeast recombinant DNA work.

sickle-cell trait, sickle-cell anemia   The heterozygous and homozygous phenotypes, respectively, for the Hb^S allele in the gene for the β chain of hemoglobin. The tendency of the mutant hemoglobin to precipitate in venous blood give blood cells a distorted (often sickle-cell) shape.

sigma replication   *See* rolling-circle replication.

sister chromatids   Chromatids that are the two products of a chromosome replication. Sister chromatids are attached at their centromeres. In G_2, all chromosomes are represented by a pair of sister chromatids.

SOS repair   Error-prone repair of pyrimidine dimers and other gross DNA damage, involving imprecise polymerization.

Southern blot   A technique by which DNA fragments, separated by electrophoresis, are transferred from a gel to a nylon membrane, for later probing with a nucleic acid probe.

specialized transduction   *See* transduction.

splicing   The removal of introns (q.v.) during mRNA processing.

staggered cut   An endonuclease cut in a DNA duplex in which the bonds broken in the two chains are offset by a defined, small number of nucleotides.

strand   A term used to designate a chromatid or chromosome, each of which contain a DNA duplex. The term is also used at the molecular level to designate a single nucleotide chain of a DNA duplex.

structural gene   A gene encoding an enzyme protein or a protein of cell structure, as distinct from a regulatory gene encoding a protein that controls gene expression.

subunit   In proteins, subunits are the component polypeptides.

supercoiling   Applied to DNA, further twisting of the double helix to impose a higher level of coiling on the molecule.

superinfecting   Designates a phage that infects a lysogen that already carries a prophage.

suppressor   A mutation that overcomes the phenotypic effect of another mutation.

symbiont   An organism that lives with or in another, both of them benefitting from the association. Often used to designate bacteria or viruses that endow cells with new and beneficial capabilities. The term is also more loosely used to describe DNA-containing organelles.

synapsis   Chromosome pairing.

synaptonemal complex   A proteinaceous structure found in its most developed form during Pachytene. It binds homologs together at this stage, and is the site of chiasma formation.

synonymy   A property of the genetic code, in which more than one codon can specify a given amino acid.

tandem duplication   Duplicate segments of DNA lying next to one another, in direct or reverse orientation.

telomere   A special repeated DNA sequence at the ends of eukaryotic chromosomes.

telophase   The last stage of mitosis or meiosis, during which chromosomes become enclosed by a nuclear membrane, and loose their cytological distinctness.

temperate phage   A bacteriophage capable of lysogeny (becoming a prophage in the host).

temperature-sensitive   Conditional mutants that do not display a phenotype at a lower temperature, but do so at a higher temperature. Mutations imparting a phenotype only at low temperatures are called cold-sensitive.

template   In RNA or DNA synthesis, the strand used to direct the synthesis of a new nucleotide strand.

tertiary structure   The overall, three-dimensional shape of a macromolecule, particularly a polypeptide.

testcross   A cross of an organism of unknown genotype to a mate that is homozygous recessive for any gene of interest.

tetrad   The four meiotic products of a single meiotic cell.

tetraploid   A cell or organism with four sets of chromosomes (4N).

tetratype Speaking of two genes, a tetrad with two parental and two recombinant meiotic products.

theta replication Replication of a circular chromosome, plasmid or phage to form two circles, by uni- or bidirectional movement of replication fork(s) around the original molecule. Distinct from rolling-circle replication (q.v.).

three-point cross A cross in which the parents differ by alleles of three genes of interest.

threshold trait A trait determined by numerous genes, but which falls into only a few phenotypic classes, often simply the presence or absence of a characteristic. The threshold is the point at which the genotype has enough of the alleles promoting the trait that the trait gains expression.

thymidine dimer The most common type of pyrimidine dimer. A covalently linked, adjacent pair of thymine bases on a single nucleotide chain. Caused by ultraviolet light treatment of DNA.

time-of-entry method A method of mapping bacterial genes in which the times markers enter F^- cells from Hfr cells is measured by periodic agitation of the culture and plating of exconjugants.

trans The arrangement of two genes in repulsion (q.v.).

trans-activating factor A regulatory protein, often with a binding site on the DNA at the 5′ end of the gene it regulates.

transcript An mRNA molecule transcibed from DNA. A primary transcript is the RNA molecule before it is processed. *See* RNA processing.

transcription Formation of an RNA molecule by RNA polymerase, using DNA as a template. Applies to formation of mRNA, rRNA and tRNA.

transduction Transfer of bacterial genes from donor to recipient bacteria by means of phage coats. **Generalized transduction** involves packaging only bacterial DNA, often randomly, into phage coats. **Specialized transduction** involves transfer of bacterial genes attached to a phage DNA, owing to their proximity in the lysogenic bacterium.

transfer RNA (tRNA) The short, folded mRNA molecules which have anticodons that can pair specifically with codons of mRNA, and to which corresponding amino acids can be attached prior to their use in translation.

transformation Uptake of DNA and either integration into the chromosome of the recipient or autonomous replication of the DNA within the recipient.

transition A base-pair substitution in which purine is substituted for purine, and pyrimidine for pyrimidine. *See* transversion.

translation Protein synthesis. Specifically, the decoding of mRNA into the corresponding amino acid sequence.

translational control Regulation of gene activity by control of the rate of translation of an mRNA.

translocation A genetic change in which a chromosome segment is attached to a new location in the genome. *See* reciprocal translocation.

transposable element, transposon DNA elements with inverted terminal repeat sequences, capable of transposing their position within the DNA of the same cell. Transposable element is the general term for all such DNAs. The smallest are **insertion elements,** encoding at most the enzyme (**transposase**) required for transposition. Larger, (**compound) transposons** typically bear genes of identifiable function such as drug resistance.

transposition The movement of a mobile DNA element. *See* transposable element.

transversion A mutation in which purine is substituted for pyrimidine and vice versa. *See* transition.

triplet Three nucleotides in sequence in a nucleotide chain. Often used as a synonym for codon.

triploid Having three sets of chromosomes (3N).

trisomy Aneuploid condition in which an otherwise diploid nucleus has three homologs of one type of chromosome (2N + 1).

Turner syndrome A human female phenotype characterized by sterility and disturbances of secondary sex characteristics conferred by the XO genotype.

uORF An upstream open reading frame. Sometimes found in the 5′ leader of mRNAs and associated with regulation of translation or (in the case of attenuation) the continuation of transcription.

unequal crossing over Abnormal recombination event which is not exactly reciprocal: one product is larger, the other smaller by the same amount. Often seen after the improper pairing of homologs with tandem duplications.

uniparental Applied to inheritance, implies asexual reproduction. Not used, as a rule, to designate parthenogenesis.

utilization mutant A mutant that differs from the standard by being able or unable to use a particular compound added to the medium for growth.

vector   In gene cloning, a plasmid or other replicating entity in which DNA fragments of interest ("inserts") can be propagated in bacterial hosts.

virion   A mature virus or bacteriophage particle.

virulent phage   A phage species that displays only lytic infections.

wild type   The genotype or phenotype taken as the standard laboratory stock for genetic work.

X : A ratio   The ratio of the number of X chromosomes to the number of *sets* of auto-somes. The critical determinant of sex in *Drosophila*.

xeroderma pigmentosum   A hereditary condition in which the bearer has great sensitivity to UV light, which causes many cancerous lesions to form in the skin. (Generally due to impairment of a DNA repair mechanism.)

zygote   The diploid cell resulting from fusion of two haploid gametes (fertilization).

zygotene   The second stage of Prophase I of meiosis, during which homologs accomplish synapsis. Synonym: synaptene.

Index